PHOTOGRAPH TAKEN BY ROCKET–CAMERA OVER MEXICO

This shows the curvature of the earth seen from an altitude of 100 miles. The dark patch is the Gulf of California. The width of the area photographed is approximately 250 miles.   (*Naval Research Laboratory, Washington, D.C.*)

Frontispiece

THE NATURE OF
THE UNIVERSE

A Series of Broadcast Lectures

BY

FRED HOYLE
Fellow of S. John's College, Cambridge

BASIL BLACKWELL
OXFORD
1950

First printed, April, 1950
Seventh impression, October, 1950

Printed in Great Britain for BASIL BLACKWELL & MOTT LIMITED
by A. R. MOWBRAY & CO. LIMITED, London and Oxford

PREFACE

IN preparing these lectures for publication I have decided to follow the original broadcast scripts as closely as possible. There is so much difference between the written and the spoken word that any attempt to cast these talks into a literary mould would have involved a major task of reconstruction. Wherever I have felt additional information to be desirable I have added a note at the end of the book: so if some point in the text should be found puzzling, it may be that a better understanding can be obtained by consulting one of these supplementary notes.

It is not too much to say that but for the constant help and encouragement of my wife the original broadcast lectures would never have been given. I am also deeply indebted to Mr. Peter Laslett, whose sure instinct for the right way 'to put an idea across' was in a large measure responsible for the degree of clarity achieved. Many of the most graphic remarks and phrases were suggested to me by Mr. Laslett.

F. H.

ST. JOHN'S COLLEGE
    CAMBRIDGE
April 3, 1950

CONTENTS

THE NATURE OF THE UNIVERSE

I

THE EARTH AND NEARBY SPACE

> Doubt thou the stars are fire,
> Doubt thou the sun doth move,
> Doubt truth to be a liar,
> But never doubt I love.

THUS Hamlet wrote to Ophelia, appealing twice to astronomical matters in one verse. For Shakespeare lived in a day when cosmology meant much to the average man. So it is not surprising to find him using cosmological ideas and imagery, one might almost say on every possible occasion. To the Elizabethans a realization of the size of the Earth and of the nature of nearby space was exciting news; it destroyed once and for all the tight little cabbage-patch world in which man had lived throughout the medieval age.

During the following three centuries, although cosmology came to mean more to the mathematicians and astronomers, it had less and less effect on the outlook of people in general. But in recent times—and by recent times I mean the last fifty years or so—it has become increasingly plain that the stage is being set for the next cosmological revolution in our way of thinking. The popularity of the well-known books of Eddington and Jeans was an obvious sign of reawakening interest in the relation of man to the universe as a whole.

What the B.B.C. has asked me to do in these talks is to describe the changes that have taken place in cosmology since the work of Jeans and Eddington. That is to say, to put before you the new developments of the last ten years or so. In some matters this means bringing up ideas that are either current or only two or three years old. I shall also have to discuss various issues that are still controversial. So it is important for you to realize that there is no present finality to the discussion of many of the questions that will turn up, and that future work is certain to throw more light on almost every problem.

My general plan, which it may be useful for you to know in advance, is to proceed outwards from ourselves and to begin with the Earth and its immediate surroundings. Next week I shall try to describe what we know of the inner workings of the Sun and similar stars, and to discuss whether energy can be generated on the Earth by processes analogous to those occurring in the Sun. Then in the third talk we move on to matters of wider scope, both in space and time; to deal with the way the stars of the Milky Way are arranged in space, and how they have originated and what their ages are. After that I shall show how the most violent explosions in nature are related to the origin of the Earth itself. In the final lecture we shall be concerned with the Universe as a whole, especially with its expansion and with the question of how it is created, while at the end I shall say how, in my opinion, the conclusions of the New Cosmology affect our philosophical and religious outlook.

Before starting the main astronomical discussion there are one or two further points of a general nature that I should like to cover. For a hundred years after the death of Newton a thorough inquiry into the nature of the Universe was still believed to be impossible. We know now that this belief is not correct, but throughout the eighteenth century there were apparently very good reasons for it. The astronomer is severely handicapped as compared with other scientists. He is forced into a comparatively passive role. He cannot invent his own experiments as the physicist, the chemist, or the biologist can. He cannot travel about the Universe examining the items that interest him. He cannot, for example, skin a star like an onion to see how it works inside.

Yet in spite of these obvious difficulties progress made during the nineteenth century showed that the situation was not so hopeless as it once had seemed. And with the coming of the twentieth century the astronomical scene has changed to a degree that could hardly have been thought possible by the early investigators. Instead of our being, as it were, small boys at the holes in the circus tent struggling to get even an imperfect peep at the great show, we now see that we have really got ringside seats from which the Universe may be observed in all the majesty of its evolution. This transformation has arisen mainly from the work of the American observers, who have exploited with the greatest skill and imagination the large telescopes that can be used in their favourable climate, like the ones at Mount Wilson. Just as a blazing bonfire is to a penny candle, so is the observational progress achieved in the last few decades to the work that came before.

But observation with the telescope is not sufficient by itself. Observation tells us, for instance, that while the majority of stars are common or garden specimens like the Sun, others are so brilliant that they act as beacons shining from the depths of space. Still others, although no larger in size than the Earth, are so intensely hot at their surfaces that they emit the short wavelength radiation known as X-rays instead of ordinary light and heat. Finally, at the other extreme of size some stars are so huge that if the Sun were replaced by one of them it would fill up most of the space occupied by the solar system; that is to say, the Earth would find itself lying deep inside the gigantic body of the star. But observation alone will not tell us why there are these various varieties of star nor what are the connections between them. Nor does observation tell us of its own accord how the stars have come into being, or what their ages are, or what will ultimately happen to them. To answer these questions and many others like them we have to enter the province of theoretical astronomy.

If I were asked to define theoretical astronomy in one sentence I should say that it consists in discovering the properties of matter, partly by experiments carried out on the Earth and partly through the detailed observation of nearby space, and in then applying the results to the Universe as a whole. It may reasonably be asked whether this is a valid procedure and perhaps I had better deal with this question before we go any further. Can we expect the information obtained from one particular small region of space and over one particular small range of time to be applicable at all times throughout all space? For example, did matter in a remote star some 1,000,000,000 years ago behave in basically the same way as matter does on the Earth at the present time? Questions of this kind constitute one of the central issues of Einstein's theory of relativity. For the principle of relativity is simply a statement that our local results do indeed have universal validity. In short if relativity is correct then our general procedure in theoretical astronomy is guaranteed.

PRINCIPLE OF RELATIVITY

I should like to expand this a little even though the point I want to make may seem rather difficult. The procedure in all branches of physical science, whether in Newton's theory of gravitation, Maxwell's theory of electromagnetism, Einstein's theory of relativity, or the quantum theory, is at root the same. It consists of two steps. The first is to guess by some sort of inspiration a set of mathematical equations. The second step is to associate the symbols used in the equations with measurable physical quantities. Then the connections that are observed to occur between various physical quantities can be obtained theoretically as the solutions of the mathematical equations. This process has two important advantages. It not only makes it possible to condense an enormously complicated mass of experimental information into a few comparatively simple equations, but it also brings out new and previously unsuspected relations between the physical quantities. What Einstein's principle of relativity states is that wherever you are in the Universe, whatever your environment, the same mathematical equations will suffice to describe your observations. I hope that you will agree that this is a very powerful statement.

But is there any direct proof of the principle of relativity? Unfortunately no, for it is a characteristic of scientific method that there can be no proof of this. Then why call it the '*principle* of relativity' instead of the postulate of relativity? Here I must admit that when scientists make categorical statements, that in their nature cannot be proved, it is usual to refer to them as 'laws' or 'principles,' perhaps only with the object of dissuading the uninitiated from asking awkward questions.

But please don't imagine that because of this the principle of relativity is simply a piece of guesswork on a par with backing a horse or making a stab at a football pool. It has been found possible, for instance, notably by Professor Dirac of Cambridge, to deduce new and important conclusions and in every case the results derived from the principle of relativity are found to agree with observation. In other words, all experience up to date shows that relativity works.

I wish I could give you some idea of the strength of such arguments. But here, as with many other issues in these talks, it is necessary to go deeply into the mathematical investigations if their real force is to be fully appreciated, and this is just what the B.B.C. has conjured me not to do. Even so I should like to give you one example of the use of analytical thought. A very fine problem was invented in 1938 by Sir Arthur Eddington. This problem, which is not nearly so well known as it deserves to be, is specially remarkable, for although it needs considerable analytical power to solve it, yet no more actual mathematics is involved than is possessed by the average child of ten. In short it consists of reconstructing the whole of an innings in a cricket match from the score card.[1] All that is needed

[1] See Appendix on p. 22.

is the batsmen's scores, the bowling analysis, and a few items of information such as who opened the bowling and who received the first ball. From these particulars it can be deduced who bowled every ball and what happened—the runs scored, the wickets taken, and so on. Do try it if you are interested in such things. I am sure it will give you a very lively idea of the effectiveness of the sort of mathematical argument that underlies many of the topics I shall be considering.

But it is now high time that we came to the astronomy and there is clearly no better place at which to begin our survey than with the Earth. The surface of the Earth is very nearly a sphere of about 8,000 miles diameter. The inside of the Earth consists of a central core rather more than 3,000 miles in diameter, surrounded by a thick rocky shell which for the most part is more rigid than steel. The essential features of the core, which have been established through the study of earthquake waves, are these: it contains fluid, and its density is substantially higher than the rocks composing the shell. At present there is no general agreement on the composition of the core. What may be called the classical view is that it consists of molten metals, particularly iron. But this idea has been challenged in the last two or three years by W. H. Ramsey of Manchester, who maintains that the core consists of essentially the same rocks as the shell, the distinction between the core and the shell being not one of composition but of the physical state of the material. A decision between these two opposing opinions must be left to the future.

There is also the possibility that earlier ideas on the temperature of the Earth's interior may have to be revised. It was argued at one time, from the fact that it rapidly becomes hotter as one descends a deep mine, that the central regions of the Earth must be at a very considerable temperature, at more, say, than 3,000° C., which exceeds the boiling point of iron. Such an argument would be valid if the rise of temperature in a mine were due to an outward flow of heat from the central regions. But this is not so. The heating in a mine, and possibly also the heating that causes the outburst of a volcano, is almost wholly due to the decay of radioactive substances, which curiously enough are confined to a thin skin of surface rocks not more than about twenty miles deep. Why this should be so is a point I shall return to later. So we see that there is no direct evidence in favour of an Earth that is really hot inside. What indirect evidence there is points in the opposite direction and suggests that the centre may be no warmer than a wood fire.

It would be possible for us to discuss a whole series of similar interesting topics. For instance, there is the evidence in favour of the view that the continents are of almost permanent shapes, that they looked nearly the same 1,000,000,000 years ago as they do now. Or the evidence that the circumference of the Earth has shrunk by nearly five per cent in the last 500,000,000 years, and how Professor Jeffreys of Cambridge has used this to explain the origin of the mountains. But it is now important that we should leave the Earth, and I say this in some sense literally, for I think that within 100 years it may indeed be possible to leave the Earth, or at any rate for rockets containing radio-operated cameras to do so. When this does happen, astronomy is certain, I think, to make marked changes in our whole outlook on life; changes that are likely to be as far-reaching as those that followed the Copernican theory of the planetary motions.

THE EARTH SEEN FROM OUTSIDE

Once a photograph of the Earth, taken from outside, is available, we shall, in an emotional sense, acquire an additional dimension. The common idea of motion is an essentially two-dimensional idea. It concerns only transportation from one place on the surface of the Earth to another. How many of us realize that but for a few miles of atmosphere above our heads we should be frozen as hard as a board every night? Apart from the petty motion of the aeroplane, motion upwards as yet means nothing to us. But once let the possibility of outward motion become as clear to the average man at a football match as it is to the scientist in his laboratory, once let the sheer isolation of the Earth become plain to every man whatever his nationality or creed, and a new idea as powerful as any in history will be let loose. And I think this not so distant development may well be for good, as it must increasingly have the effect of exposing the futility of nationalistic strife. It is in just such a way that the New Cosmology may come to affect the whole organization of society.

Now what does the contemporary astronomer expect such a photograph, a colour photograph, of the Earth to look like? There will be brilliant white patches where the sun's light is reflected from clouds and snow-fields. The arctic and antarctic will on the whole appear brighter than the temperate zones and the tropics. There will be all shades of green, varying from the light green of young crops to the sombre darkness of the great northern forests. The deserts will show a dusky red, and the oceans will appear as huge areas that look grimly black, except occasionally they will be illuminated by a blinding flash where conditions allow the Sun's light to be powerfully reflected, much as we sometimes see a brilliant shaft of sunlight reflected from the windows of a distant house. The whole spectacle of the Earth would very likely appear to an inter-planetary traveller as more magnificent than any of the other planets.

So much then for the Earth. Perhaps we should next take a look at our nearest neighbour, the Moon, which is a ball of rock only about an eightieth as massive as the Earth. The Moon is a satellite; that is to say, it moves along a nearly circular path around the Earth. The time required for it to go once around this path is called the lunar month, which is at present about twenty-seven days. I cannot explain in this talk exactly why it should be so, but owing to the effect of the tides our satellite is getting steadily further away from us and the lunar month is getting steadily longer. If we go back into the past the converse is true; namely, that the Moon was then nearer to us than it is now. If we worked backwards in time as far as the birth of the Earth, which, as I shall show in my third talk, occurred about 2,500,000,000 years ago, it is probable that the Moon was then very close to the Earth, if indeed the two were not in actual contact. During the aeons that have since elapsed the tides have not only caused the recession of the Moon, but its rotation has been gradually slowed down. It turned more and more slowly until now it keeps one face permanently towards us. No one has yet seen the opposite side of the Moon.

CRATERS ON THE MOON
A photograph showing the area Licetus to Theophilus
(*Mount Wilson and Palomar Observatories*)

The Moon has no detectable atmosphere, and its surface is severely pockmarked, looking as if it had been bombarded by a host of large celestial missiles. And this is very likely just what has happened. At one time it was thought that the lunar craters were extinct volcanoes, but for the following reasons this now seems unlikely. Some of the craters are over a hundred miles across, and the big ones show almost the same structure as the small ones. Terrestrial volcanic craters, on the other hand, are only a few miles across and do not show the same uniformity of structure. Besides, it is certain that volcanic activity on the Moon is quite negligible at the present time.

Perhaps the strongest argument in favour of the bombardment theory is that the amount of material in the walls surrounding a crater can actually be estimated, and it turns out to be just the amount required to fill in the hole in the floor of the crater. But in spite of this patent clue the bombardment theory has not gained general currency among astronomers because it was thought that a crushing argument could be brought against it. There are large areas of the Moon where no craters can be found. How have all the missiles contrived to miss these areas, whereas in other places the craters are almost overlapping each other? Only the other day the way round this apparent difficulty was pointed out to me by my colleague, Gold. The fierce heating of the lunar surface rocks by day and the cooling by night must lead to an alternate contraction and expansion which causes small bits of rock to flake away from the surface. These particles of dust tend to work their way to the lower parts of the Moon where they have accumulated as gigantic drifts that cover the

B

underlying craters. I think that this brand new idea is almost certainly correct, because it not only overcomes the old objection, but it also explains those cases where the walls, or a portion of the walls, of a crater stick straight out of an apparently flat plain. These are simply the cases where the drift of dust is not sufficiently deep to cover the craters entirely.

I said a few moments ago that no one has as yet seen the other side of the Moon. But we might live to do so. If to-night a chunk of material the size of a mountain should strike the Moon obliquely, there is no doubt that the Moon would be set in rotation again, and the unseen side would turn slowly towards us. As you will guess, this is not very likely to happen, but there is nothing impossible about it.

This brings me back to the Earth for a minute. The Earth must have suffered an even greater bombardment than the Moon. But with the exception of one or two recent formations, such as the famous meteor pit in Arizona, the resulting craters on the Earth have been removed through erosion by wind and water. Do missiles still strike the Earth? Well, a piece of rock, probably about the size of a house, hit Siberia in 1908, and the resulting blast is said to have felled trees over a wide area. At any moment another celestial cannon-ball may hit the Earth anywhere. For all we know, the whole of London may be wiped out during the next five minutes. But there is no occasion for alarm, the odds against it are very large.

Pride of place in our sky belongs to the Sun, not the Moon. The Sun is about 360 times further away from us than the Moon and is about 300,000 times as massive as the Earth. The Sun shines by its own light, whereas the light from the Moon, and the light that would be seen from the Earth if we could observe it from a distance, is simply reflected sunlight. I don't wish to say much about the Sun to-night as it forms the main topic of my next talk. But there is one effect of the Sun—namely, its control over the motion of the Earth—that is so important that I feel something must be said about it, even though this means spending the next few minutes discussing the history of astronomy.

The direction of the line joining the Sun to the Earth is observed to change from day to day. Starting at a particular time, this line swings round in a plane and comes back to its original position in a year. The yearly rotation of this line can equally well be regarded as arising from a motion of the Sun around the Earth or alternatively as a motion of the Earth around the Sun. From a geometrical point of view these two pictures are equally simple. But as soon as we also consider the motions of the other planets this is no longer the case. A planet is a body, small in size compared with the Sun and near to us compared with the stars, that shines by reflected sunlight, as the Moon does. In ancient times five planets were known, other than the Earth. They were recognized from their motions on the sky which have the effect of constantly altering their positions relative to the stars. It was found that they all lie nearly on the plane swept out in the course of the year by the line joining the Earth and the Sun, so that the whole solar system forms a sort of flat plate-like structure.

Now the paths traced on the sky by the planets are not simple like the steady circular motion of the Sun. The first attempts to explain these paths in terms of motions in space were made on the assumption that it is the Earth that is fixed and that the planets as well as the Sun move in orbits round the Earth. Although much ingenuity was displayed by such men as Eudoxus and Ptolemy, the picture they drew was complicated, and as more observational data came to hand it became more so, until it was a thorough mess. As far as we know the first man to perceive that a far simpler description could be achieved by taking the Sun as the centre of the system was Aristarchus of Samos, who lived in the third century B.C. He found it possible to explain the observations by supposing that all the planets, the Earth included, move around the Sun in essentially circular orbits of various radii. If sufficiently detailed historical records were available it would be an interesting study in prejudice to see why Aristarchus' views were ignored by his fellow Greeks. At all events they were forgotten until revived by Copernicus nearly 2,000 years later.

The conflict between the Copernican theory and the Roman Catholic Church is well known, especially the part played by Galileo. In the course of his Third Programme lectures on 'Christianity and History' Professor Herbert Butterfield offered a mild apologia for the attitude of the Church. Much as I enjoyed Professor Butterfield's lectures, I am afraid I am out of sympathy with him on this point. I regard the statements of Galileo's opponents as exemplifying the dictum of the humorist Josh Billings, which ran something as follows: 'It ain't what a man don't know as makes him a fool, but what he does know as ain't so.' The case for the Copernican theory is not that it is right or true in some absolute sense, but that it was the only point of view from which progress could have been made at that time. In short, that it had the virtue of simplicity, and this was demonstrated with great cogency and skill by Galileo.

How far Galileo was justified in advocating the Copernican theory is shown by subsequent events. Astronomy now began to advance in giant strides. Kepler found that the planets do not move exactly in circles but in nearly circular ellipses. The notion of gravitational attraction occurred to many people, including Halley, Wren, Hooke, and Newton in this country. By considering the planetary orbits to be circles a quantitative formulation of gravitation was obtained. But could this formulation explain the ellipses found by Kepler? Only Newton was able to prove this, and in doing so he set the pattern for all subsequent scientific investigations. But Newton did not stop there. He went on to show that many other apparently disconnected observations—the tides, for example—could also be explained by gravitation. His work was on such a colossal scale that it is natural to find the following century and a half given over to the consolidation of his ideas. The detailed explanation, in terms of gravitation, of every detail of the planetary motions became the chief programme of mathematicians and astronomers. It was Adams and Le Verrier who closed this chapter in the history of science by using the Newtonian theory to predict the existence and the position of a new planet named, subsequent to its observational detection in 1846, Neptune.

From now on astronomy takes a different turn. The motions of the planets cease to be of chief concern. The nature of the Sun and the stars, and latterly nothing less than the whole Universe, provide the subjects to be attacked by scientists. I shall say nothing more of this to-night because I shall be dealing with these topics later in this series. But I should like to take a more detailed look at the planetary system now. Suppose we make a plan of how the Sun and the planets are arranged. In our plan let us represent the Sun as a ball six inches in diameter, the sort of thing you could easily hold in one hand. This, by the way, is a reduction in scale of nearly 10,000,000,000. Now how far away are the planets from our ball? Not a few feet or one or two yards, as many people seem to imagine in their sub-conscious picture of the solar system, but very much more. Mercury is about 7 yards away, Venus about 13 yards away, the Earth 18 yards away, Mars 27 yards, Jupiter 90 yards, Saturn 170 yards, Uranus about 350 yards, Neptune 540 yards, and Pluto 710 yards. On this scale the Earth is represented by a speck of dust and the nearest stars are about 2,000 miles away. You may wonder how the various distances in the solar system have been established. In principle the methods used are only a matter of elementary trigonometry, but in practice very intricate measurements are necessary and even then extensive calculations have to be made in order to extract the required information. The present Astronomer Royal, Sir Harold Spencer Jones, established the current values about ten years ago.

The wide spacing of the planets is even more remarkable than our plan would suggest because the biggest planets are the ones lying at great distances from the Sun. Venus is slightly smaller and Mercury and Mars appreciably smaller than the Earth. Jupiter, on the other hand, is more than 300 times as massive as the Earth, but even so Jupiter is still much smaller than the Sun. Indeed, if the Sun were suitably divided up it would make more than a thousand bodies of the size of Jupiter. Saturn is about 95 times more massive than the Earth, Uranus about 15 times, and Neptune 17 times. The exact size of Pluto is not known, but it is certain that Pluto is small like Mercury, Mars, and Venus. The four large planets, Jupiter, Saturn, Uranus, and Neptune, are often referred to as the great planets, and they lie at distances from the Sun that, on our plan, exceed 90 yards. I have been insistent on getting the scale of the planetary system right, because an appreciation of the wide spacing will be of great help when later we come to consider how the planets originated.

Let us leave the solar system for the present by taking a more detailed look at some of its members. Venus might be called the twin sister of the Earth, and therefore naturally claims our attention first. Venus is not very aptly named, for she modestly hides her surface by a perpetual bank of white cloud. The nature of this cloud is a bit of a mystery. At first sight it might be thought to consist of water drops, as the clouds on Earth do. But this cannot be correct, otherwise the presence of water vapour would have been detected. So far the only gas found in the atmosphere of Venus is carbon dioxide, which is present in enormous quantities. All attempts to detect substances with cloud-forming properties similar to water have failed.

A possible solution of the problem is suggested by the work of the French astronomer, Lyot, who discovered only a few years ago that the sunlight scattered by the clouds of Venus has the characteristics of light scattered from fine white dust. The intense heating that occurs on the sunlit face of the planet must cause small particles to flake off the surface rocks just as in the case of the Moon, and it seems possible that great clouds of these particles rise upwards in the carbon dioxide atmosphere in a sort of gigantic fountain and are then convected round to the dark side of the planet. If this idea is correct any future traveller to Venus will be rewarded by a fine spectacle.

There is another mystery about Venus. It takes her more than twenty days to rotate on her axis. As it seems probable that all the planets at their birth had rotation periods of about ten hours, it is an interesting question as to what process has slowed down the rotation of Venus to such a marked extent. I cannot enter now into any detail beyond saying that one possibility is that the Sun exerts a huge tidal influence on the shifting drifts of dust that probably form the surface of Venus, and that the only other possibility known to me is a suggestion by Lyttleton of Cambridge, that Mercury may once have been a satellite of Venus.

The most interesting feature connected with the planet Mercury concerns its orbital motion around the Sun. The ellipse in which Mercury moves is more flattened than the orbit of any other planet, and the strange discovery made about seventy-five years ago by the famous French mathematician, Le Verrier, was that the longest axis of the ellipse is slowly turning in space. At first this was thought to be an effect due to a new planet moving in an orbit still nearer to the Sun. So sure were astronomers of the existence of this hypothetical planet, especially after their experience concerning the discovery of Neptune, that they had the name Vulcan ready to be attached to it as soon as its existence was confirmed by observation. But it is curious how rarely in science the same tactics can be repeated with success. Whereas Neptune had been detected almost immediately once the predictions of Adams and Le Verrier were made known, the observers now searched in vain for Vulcan. It is also curious that Le Verrier's failure to find a new planet in his second attempt should turn out to be of far greater scientific importance than success would have been. For success would only have added a small body to the solar system, whereas failure meant that at long last a flaw in the Newtonian theory of gravitation had been found. About forty years later the resolution of this issue was to form one of the cornerstones of Einstein's general theory of relativity.

Now let us move outwards from the Earth in a direction away from the Sun. What can we expect to find on the first planet, Mars? Much very interesting information has been obtained in the last two or three years by Kuiper of the Yerkes Observatory near Chicago, who has used an ingenious new technique to show that the white polar caps of Mars are indeed composed of snow. The atmosphere contains more carbon dioxide than that of the Earth, and there may also be nitrogen. The thin clouds responsible for the haze so troublesome to the observer are probably composed of ice crystals. No oxygen can be detected, though free oxygen may once have existed on Mars in appreciable quantities. There are green markings on the surface of Mars that seem to vary from time to time. The changes in these markings may well be due to the growth and decay of plants similar to the rock lichens found on the Earth. It is not surprising that plant life on Mars should be confined to such a low form, for lichens require little moisture and they can survive at lower temperatures than the more usual terrestrial plants. Both these properties would be important requirements on Mars. Are there also other forms of life? I do not think any other answer can be given to this question but to say that we must wait and see.

Beyond Mars lie the great planets, Jupiter, Saturn, Uranus, and Neptune. They are well furnished with satellites; Jupiter has at least eleven, Saturn nine, Uranus four, and Neptune two. Their atmospheres are about as different from that of the Earth as you could imagine, being composed of methane, ammonia, and probably hydrogen. An old fallacy, exposed about twenty years ago by Jeffreys, was that on account of their large sizes the surfaces of the great planets must be hot. Observation confirms Jeffreys' argument and shows that their temperatures are indeed lower than about $-150°$ C. Below a thin fringe of atmosphere, it is probable that in each of these planets there are thick shells of ice and solid hydrogen overlying an inner ball of rock and metal. There are many questions that we should like to be able to answer. What is the red spot that changes so markedly from time to time in the atmosphere of Jupiter? Is it a solid of some sort floating in a sea of gas? What are the particles that form the rings of Saturn? Are they indeed small ice crystals? Do the great planets have large magnetic fields? Are the five inner satellites of Saturn really gigantic snowballs? It would be possible to go on for a long time asking this sort of question, and, interesting as their discussion might be, there would still remain the larger problem of the origin of the planets themselves. But as this issue arises again in a later talk, I will make an end by asking whether the fragment of space that we have considered here has anything exceptional to distinguish it from all other parts of the Universe. Is this procession of night and day, this movement of the Earth and planets around the warming Sun, something really special, or are there lots of places where similar systems occur? When you look at the heavens, how many of the stars you see have planets encircling them and on how many of these planets might living creatures look out on a very similar scene?

To give a numerical estimate I would say that rather more than a million stars in the Milky Way possess planets on which you might live without undue discomfort. If you were suddenly transported to one of them you would no doubt find many important changes, but the changes would not be as remarkable as the similarities. And if you had been brought up from birth on another planet moving around another star you would feel the same sort of emotional attachment to it as we feel for the Earth. You would look out on the Milky Way and wonder, as we do, what proportion of the stars also have attendant planets with living creatures on them. One of the comparatively insignificant stars that you would see would be the Sun. But, even with a powerful telescope, you would not see the Earth or any of the planets of the Sun's system, for they are far too faint to be seen at immense distances. In short, you would have to infer our existence just as we have to infer yours.

APPENDIX

THE following is the problem set by Sir Arthur Eddington (which appeared in Caliban's column in *The New Statesman*) and referred to on page 6.

Extract from the score of a cricket match between Eastershire and Westershire:

EASTERSHIRE—Second Innings

A. A. Atkins - - - -	6
Bodkins - - - -	8
D. D. Dawkins - - -	6
Hawkins - - - -	6
Jenkins (J.) - - -	5
Larkins - - - -	4
Meakins - - - -	7
Hon. P. P. Perkins - - -	11
Capt. S. S. Simkins - - -	6
Tomkins - - - -	0
Wilkins - - - -	1
Extras	0
Total	60

Bowling Analysis:

	Overs	Mdns.	Runs	Wkts.
Pitchwell	12.1	2	14	8
Speedwell	6	0	15	1
Tosswell	7	5	31	1

The score was composed entirely of singles and fours. There were no catches, no-balls, or short runs.

Speedwell and Tosswell each had only one spell of bowling. Pitchwell bowled the first over, Mr. Atkins taking first ball. Speedwell was the other opening bowler.

Whose wickets were taken by Speedwell and Tosswell?

Who was not out?

What was the score at the fall of each wicket?

SOLUTION

Since Pitchwell bowled the 1st over and the unfinished 26th over, he must have changed ends. Thus when Speedwell came off at the end of the 12th over, Tosswell bowled the 13th, Pitchwell the 14th, etc.

Tosswell's 5 maidens show that 31 runs were hit off him in 2 overs. The strokes were, therefore, 7 fours and 3 singles. Not more than 5 batsmen can have been concerned in them (3 in the over in which Tosswell took a wicket, and 2 in the other over). Thus 2 batsmen scored 2 fours; these were Bodkins and Perkins, no one else having reached 8.

Since Bodkins' 8 was hit off Tosswell, he played the first 12 overs without scoring. We can therefore write down at once the scores at the fall of the first few wickets, namely, 6, 12, 18, 23, . . ., provided in each case that the wicket fell before the 13th over. Pitchwell was bowling to Bodkins at the odd scores; thus wickets taken by him fell at even scores. The fourth wicket (Jenkins) which fell at an odd score was taken by Speedwell. The 13th over cannot begin before Speedwell has secured his one wicket.

Whilst Bodkins is not scoring no over can include more than one single. Therefore Speedwell's 6 overs, none of which were maidens, consist of 3 containing a four (*f*-overs), and 3 containing a single (*s*-overs). After an *f*-over Pitchwell bowls a maiden to Bodkins; since he bowled only 2 maidens, the third *f*-over must have been the 12th over. The fourth wicket fell after the four in the 12th over was scored; for if it fell earlier, 11 (at most) of the 23 runs came from Speedwell, so that 12 came from Pitchwell; that gives Pitchwell the impossible analysis of 6 overs, 0 maidens, 2 runs for his last 6 overs.

The 13th over opens with Bodkins (0) facing Tosswell, and Larkins (0) at the other end. Pitchwell's analysis to date is:

6 overs, 2 maidens, 8 runs, 3 wickets.

Since he has no more maidens, he gives away a single in each of his last 6 overs. Thus including the 3 singles off Tosswell (already noted) there are 9 singles to come. This is equal to the minimum number of singles contained in the scores of Meakins, Perkins, Simkins, and Wilkins; thus Larkins' score was a four (not 4 singles).

If Tosswell began with a maiden, the next over was bowled by Pitchwell to Larkins, and the condition that Pitchwell's over contains a single cannot be satisfied. Therefore Tosswell began with an erratic over, which is easily seen to be 4, 4, w, 4, 1, 4.

In the 14th over, Meakins (5) faces Pitchwell, with Larkins (4) at the other end. Neither can take part in Tosswell's second erratic over, since the batsmen then in scored fours off him. We must, therefore, fill up with maidens from Tosswell whilst Meakins slowly takes his score to 7, and gets out. This takes us to the 22nd over. The 23rd is Tosswell's second erratic over; and the rest of the innings is easily traced.

The following table shows the scoring strokes and wickets taken in each over:

Overs	Pitchwell	Speedwell	Overs	Tosswell	Pitchwell
1—2	4, 1	1	13—14	4, 4, w, 4, 1, 4	1
3—4	w, 1	4	15—16	M	w, 1
5—6	M	1	17—18	M	1
7—8	w, 1	4	19—20	M	1
9—10	M	1	21—22	M	w, 1
11—12	w, 1	4, w	23—24	1, 4, 4, 1, 4	w, w, 1
			25—26	M	w

There are alternative possibilities (not affecting the answer) in the first 9 overs; the 23rd over may have been 4, 1, 4, 4, 1. The rest of the table is unique.

Jenkins fell to Speedwell, Bodkins to Tosswell, and Wilkins was not out. Score at the fall of each wicket: 6, 12, 18, 23, 31, 41, 44, 59, 59, 60.

II

THE SUN AND THE STARS

MY main purpose to-night is to tell you what the astrophysicists have discovered recently about the inner workings of the Sun. And this will bring up their answers to a number of age-old cosmological questions. What is the Sun made of? How hot is it? Is it simply hot on its surface, or is the whole body hot, inside and outside? These are some of the things which puzzle people. Much more important is this one. What is the source of the Sun's energy? Is it growing hotter, or colder? How long will it continue to radiate light and heat at just the rate required by living creatures on the Earth? Is it getting smaller and smaller, or will it stay the same size—or even perhaps get bigger? Some of these questions, I might warn you, will take us forwards into the remote future, perhaps to a time more than 10,000,000,000 years hence.

After all this, there is another class of question, not of such wide cosmic importance but of urgent practical interest, that we must also consider. For in the study of the Sun's light and heat astronomy comes in contact with everyday affairs. Not only is sunlight a necessity for the support of life on the Earth, but it is also the ultimate source of all the energy at present used in industry. The power produced by coal and oil represents sunlight that was stored in trees and plants thousands of centuries ago. Even hydro-electric power really comes from the Sun, for it is the Sun's heat that sucks water from the oceans into the atmosphere. But falling water is not a big source of energy and it is well known that our coal and oil supplies cannot last for more than a few centuries. So it looks as if our power may finally give out, and with it the whole of our present civilization. Moreover, we need more energy, great quantities of it, if we are to go on developing at the rate we are getting used to. How are we to find a new supply of energy? Should we start growing plants with the object of trapping the Sun's light, or should we build a whole lot of miniature suns of our own? This can be done, as you know, by disintegrating uranium in an atomic pile. This will bring me to our newest, our most anxious, fear. It has been maintained by some people that an atomic explosion might fire off a nuclear chain reaction that would blow up the whole Earth. Whether this is so or not must form a part of our cosmology.

As you will realize we have a very full programme before us, so we must push on pretty hard if we are to get through to the answers to all these questions. First, then, a few general remarks about the Sun. It is the nearest of the stars—a hot self-luminous globe. Though only a star of moderate size, the Sun is enormously greater than the Earth and the other planets. It contains about 1,000 times as much material as Jupiter, the largest planet, and over 300,000 times as much as the Earth. Its gravitational attraction controls the motions of the planets, and its rays supply the energy that maintains nearly every form of activity on the surface of the Earth. There are some exceptions to this general rule: for instance, the upheaval of mountain ranges and the outbursts of volcanoes.

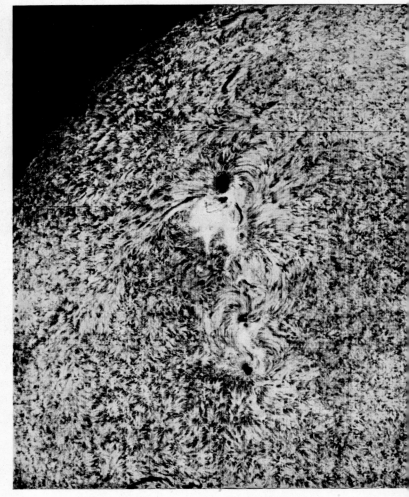

THE SUN'S ATMOSPHERE OF HYDROGEN

This photograph was obtained with the aid of the spectroheliograph, which isolates the faint light given out by the hydrogen.

(*Mount Wilson and Palomar Observatories*)

You might like to ask why the Sun is able to supply its own light, heat, and energy, whereas the Earth and the other planets only shine feebly with the aid of borrowed light? Strange as it may seem, it is best to start this problem by considering the interior of the Earth. Owing to the weight of the overlying rocks, material near the centre of the Earth is subjected to enormous pressure. Indeed, in the deep interior the pressures amount to nearly 100,000,000 lb. per square inch. It is the same inside the other planets, and in those that are larger than the Earth the pressures developed are even greater. It may surprise you that the ordinary solids and liquids of common experience should be able to withstand such terrific forces without giving way.

But if we apply this argument to the Sun we get a different answer. It can be established that, in order to withstand the weight of the overlying layers, the pressure at the centre of the Sun must be nearly 100,000 times greater than the already tremendous values occurring within the Earth. Ordinary solids and liquids certainly cannot stand up to compressional forces as great as that. If the sun were constituted like the Earth, it would collapse visibly before our eyes under the inexorable power of its own gravitational field. How then does the astrophysicist explain why the Sun does not collapse, and also why it has remained pretty much its present size, as the geologists have shown, over at least the last 500,000,000 years? There is only one possibility. The material inside the Sun must be hot, very hot, by our standards. By calculation we have discovered that near the Sun's centre the temperature must be close to 20,000,000° C. This may

C

be compared with the temperatures attained in an electric furnace, which are less than 3,000° C., or even with the surface of the Sun where the temperature is only 6,000° C. Here then we can tick off the answer to one of our original questions; namely, 'How hot is it inside the Sun?' It is 20,000,000° C., and it is very, very much hotter inside than it is at the surface.

SOLAR RADIATION

It is difficult to appreciate what a temperature of 20,000,000° C. means. If the solar surface and not the centre were as hot as this, the radiation emitted into space would be so great that the whole Earth would be vaporized within a few minutes. Indeed, this is just what would happen if some cosmic giant were to peel off the outer layers of the Sun like skinning an orange, for the tremendously hot inner regions would then be exposed. Fortunately, no such circumstance is possible, and the outer layers of the Sun provide a sort of blanket that protects us from its inner fires. Yet in spite of these blanketing layers some energy must leak through from the Sun's centre to its outer regions, and this leakage is of just the right amount to compensate for the radiation emitted by the surface into surrounding space. For if the amount leaking through were greater than the amount radiated, the surface would simply warm up until an exact balance was reached. The situation has some similarities with what happens if you heat a long metal bar at one end. Heat travels from the hotter end to the cooler end. But this analogy is not perfect. Analogies never are. Heat is carried along a metal bar by conduction, whereas in the Sun the outward leak of energy is carried by radiation. The radiation changes its character as it works its way outwards. At the surface it is ordinary light and heat, but in the central regions it takes the form of the very short wavelength radiation known as X-rays.

We now reach an important point. The rate at which radiation leaks through from the central regions and thence into outer space can be calculated—that is to say, the brightness of the Sun can be predicted theoretically. The result of the calculation depends most strongly on the amount of material present in the Sun. If, for instance, the amount were increased tenfold, the brightness would increase about a thousandfold. Not even the most enthusiastic sunbather would welcome this change, for it would cause the whole body of the Earth to melt and the rocks would bubble merrily. Then again the Sun's brightness depends on the chemical composition of its material, and also on its size. The Sun would become dimmer if it were expanded and more brilliant if it were contracted.

The first calculations along these lines were made by Eddington. In his remarkable book *The Internal Constitution of the Stars* he worked out a theoretical value for the brightness of the Sun, using as the ingredients of the calculation the quantity of material in the Sun and its known size. The theory gave a brightness nearly a hundred times too large; that is, a hundred times greater than it is known to be by observation— ordinary observation by telescope. But this was not as bad as it sounds, because Eddington had to make a guess at the chemical composition of the solar material. His first guess was that the material is predominantly composed of iron and other elements of what is called high atomic weight. The important feature of this guess was that no appreciable quantities of hydrogen and helium were thought to be present.

By about 1930, Eddington, however, had come round to the view that his original idea of a Sun made of iron was to blame for the trouble. It was found that the presence of appreciable quantities of hydrogen—the element with the lowest atomic weight of all—would make a very big difference in the theoretical result. To bring theory into line with observation, the Sun had to contain either about 35 per cent hydrogen or over 90 per cent hydrogen. Now astronomers were effectively unanimous in preferring the 35 per cent alternative, even though H. N. Russell of Princeton had shown that hydrogen is overwhelmingly predominant in the atmospheres of many stars. Here you must allow me a slight digression, for you see now the working of prejudice. Previous opinion had been that the Sun contained next to no hydrogen. When Eddington's work upset this notion it was decided to accept the lesser of two evils and the 35 per cent possibility was accordingly adopted. And this view has persisted until quite recently. A proper appreciation of the general cosmic abundance of the various chemical elements is, as we shall see, one of the most recent cosmological developments.

We reach a crucial turning point in our argument. We have seen that the interior has to be very hot indeed to prevent the Sun collapsing catastrophically. We have also seen how the rate at which the surface radiates energy into surrounding space can be calculated. I have mentioned the various items of information that constitute the basis of the calculation, and I hope that you will have noticed that at no point did

I introduce the idea that the Sun actually generates energy by nuclear transmutations taking place in its interior—the sort of thing that goes on in atomic piles. Does this mean that the brightness of the Sun is independent of whether any such energy is being produced or not? The answer to this is, yes. If the size of the Sun and the quantity and composition of the material it contains are all known, then its brightness is a fixed quantity, quite regardless of whether or not energy production occurs in the interior. This result may strike you as very surprising. Eddington's contemporaries certainly found it so. Jeans, in particular, never seems to have understood its significance.

But is nuclear transmutation taking place in the Sun nevertheless? The best way for us to make further progress in this problem is by asking how Eddington was able to deduce that energy generation must indeed be taking place inside the Sun, and at such a rate as to compensate exactly for what is being radiated into surrounding space. Let us suppose, by some magic, that we remove the sources of the solar energy. There will be no immediate change in the Sun's brightness. But as you will realize, the Sun cannot go on losing energy indefinitely without there being some important changes in its internal structure. What would the changes be? I suppose the natural answer would be to say that the Sun would cool off. But this is wrong. For, as we have seen, the inner regions could not then support the weight of the overlying layers and there would be a complete collapse of the whole body. So a cooling-off process would not be a stable one. The loss of radiant energy from the surface would lead to a very slow contraction of the whole of the Sun, and, paradoxical as it may sound, this compression would actually heat up the material. Eddington's method of determining the brightness remains valid and shows that so far from cooling off, the Sun would actually grow steadily brighter as it contracted. Calculation shows further that the reduction of the diameter of the Sun would be about a hundred yards every year. At first sight this appears to be very little—it would certainly lead to no noticeable effect, even with sensitive instruments, over the whole course of recorded history. But this is only a way of saying that the period of recorded history is extremely short. Over periods of time that are commonplace to the geologists the Sun would change a very great deal.

GENERATION OF ENERGY

If we put this argument in a slightly different form we can immediately reach our conclusion. For if throughout the geological ages some source of internal energy had not just compensated for the radiation that was being lost at the solar surface, the Sun would necessarily have shrunk by now to a tiny body. In short, it would have become much less than it is observed to be at present.

But the inference that there must be energy generation inside the Sun does not settle our difficulties. We have still to find out exactly how the energy is produced. Ordinary chemical sources are hopelessly inadequate. If, for instance, the Sun were made out of a mixture of oxygen and best quality coal, the coal would be reduced entirely to ashes in only two or three thousand years. Nor is the natural radioactivity of uranium, such as occurs in the rocks that compose the Earth's crust, sufficient to run the solar engine. Some new source depending on atomic transmutation is necessary. This requirement first made it clear to scientists that it must be possible to find nuclear processes that are very powerful sources of energy. Here, as with so many other important ideas in physics, the lead was supplied by astronomy.

How, then, is energy generated in the Sun? Two suggestions as to this were made by Jeans. One was that the Sun might contain super-radioactive material not present on the Earth, and the other that matter might even be annihilated under the physical conditions occurring in the solar interior. For various reasons that it would take me too long to describe, neither of these ideas has passed into current astrophysics. The solution of the problem lay along different lines, and, at the risk of being a little technical, I should like to go over the main developments as they occurred.

Let us transfer the scene to the interior of the atom instead of the interior of the Sun. The chemical elements are classified according to the particles contained in their central nuclei. At the lower end of the list of atoms found in nature is ordinary hydrogen with a nucleus containing one particle—a proton—while at the upper end is the commonest form of uranium, which has a complex nucleus made up of 92 protons and 146 neutrons. I hope you are familiar enough with these terms not to let them worry you.

Measurements made by Aston of Cambridge in the early nineteen-twenties showed that the best way of getting energy out of the elements at the upper end of the list is to break up the nucleus, preferably into two pieces of about the same size. As you are probably aware, the only elements for which this has so far been found practicable are uranium and thorium. Exactly the opposite situation occurs for nuclei containing less than about fifty particles. These have to be added together for energy to be obtained. Many such building-up processes are possible, but only one is of great astronomical interest. Helium is next to hydrogen at the lower end of the scale of atomic weights. If four protons could be combined so as to form an alpha-particle, as the nucleus of helium is usually called, a large amount of energy would be set free. Remembering that Eddington's work showed that the Sun must contain at least 35 per cent hydrogen, we are naturally led to ask the question: Is the conversion of hydrogen into helium the process that supplies the solar energy generation?

An important start towards answering this question was made in the early nineteen-thirties by Atkinson and Houtermans, who showed that nuclear transformation processes do indeed occur in the solar interior at roughly the required rate. The next step was taken in 1938 by Gamow and Teller, whose work may be described as bringing the ideas of Atkinson and Houtermans into line with the rapidly developing science of nuclear physics. But so far no one had earmarked the actual processes that supply the Sun's energy. This link in the chain was left to H. A. Bethe of Cornell, who showed, in 1939, that a particular set of reactions involving carbon and nitrogen as catalysts have the effect of building helium from hydrogen at just about the rate necessary to compensate for the energy radiated from the solar surface. Catalysts, you remember, are substances which help a reaction to occur but do not change themselves.

It was at this stage that my colleague, R. A. Lyttleton, and I first became interested in the problem of the structure of the Sun. It seemed to us that Bethe's work, if it were put into the calculations at the beginning instead of at the end, should lead to a considerable improvement in the whole method of investigation, which had hitherto lacked both accuracy and elegance. These troubles were due at root to the use of the observed size of the Sun as a datum of the calculations. So long as the mode of energy generation was unknown, this was a necessary procedure, but once the nuclear processes occurring in the Sun were understood, it was possible to put the whole problem in a much more direct and challenging form. Given only the amount and the composition of the solar material, is it possible to decide purely by calculation both the brightness of the Sun and what its size must be? Lyttleton and I found that this could indeed be done, and we were able to show that the results of the mathematical theory agree with observation to an accuracy of a few per cent.

This is not the end of the story. The next step leads us away from the Sun to other stars. Eddington, right from the outset, was not slow to see his theory also applied to the stars in general. His comparison of theory with observation for a group of about twenty stars was at first regarded as very encouraging, but as time went on certain discrepancies became more and more manifest. These discrepancies persisted until about two years ago, when it was realized that they can be completely resolved by a change in our view as to the chemical composition of the material composing the Sun and the stars. You will remember that in Eddington's work, consistency between theory and observation could be obtained if the Sun contained either 35 per cent or more than 90 per cent of hydrogen. The only step that was necessary to overcome the discrepancies I have just mentioned was to adopt the larger percentage. To sum up the most recent conclusions, a normal star at the time of its birth consists of about 1 per cent oxygen, nitrogen, and carbon, about 1 per cent of heavy elements such as iron, perhaps up to 5 per cent helium, and the rest hydrogen. This answers one of our original questions: What is the Sun made of? Because, as we shall see next week, the Sun is still in its infancy.

At this stage we may notice an important point relating to the origin of the planets. If the weight of hydrogen in the Sun is over a hundred times greater than the combined weight of such elements as silicon, oxygen, iron, and magnesium, it is clear that the composition of the Sun is very different from that of the planets, where the combined abundance of hydrogen is only about a third of that of the other elements. In short, there is about a three-hundredfold difference between the hydrogen content of the Sun and the planets, and this must be taken into account when in a later talk we come to consider the process that led to the formation of the planets.

By now we have implicitly answered a number of our original questions. How long will the Sun continue to radiate light and heat at just the rate required by living creatures on the Earth? Calculation shows that so long as the Sun is not seriously disturbed by processes occurring outside itself, and this matter will form one of the chief topics of my talk next week, the supply of hydrogen in the Sun will last for about 50,000,000,000 years. This does not quite answer our question, because after about 10,000,000,000 years the Sun will be getting too warm for our comfort. In other words, as more and more hydrogen gets converted into helium, the Sun will get hotter and hotter. This is another of those results that go the opposite way from what you might naturally expect. By the time the Sun has used about a third of its present store of hydrogen the climate, even at the poles of the Earth, will be getting too hot for any forms of life that at present inhabit it. At a still later stage, the Sun will become so hot that the oceans will boil and life will become extinct. So life will perish in the solar system as a whole, for the same considerations will also affect Mars, not because the Sun becomes too feeble, but because we shall be roasted.

I think this answers all our questions concerning the Sun except this one—perhaps a minor one for those whose interest in cosmology is not professional: Is the Sun going to change its size? Is it going to swell or to shrink? To deal with this, we must return to the stars. So far I have spoken as if all stars can be fitted into the scheme I have described. This is not so. Those that do have a special name: they are called the main sequence stars. There are several groups of non-conformers, and one of them, the red-giants, I want now to consider briefly. The red-giants are also of interest partly because the problem they raise is one of the most recent to receive solution, and partly because, as I shall describe next week, they serve as a clock whereby we can determine the ages of the stars.

I am rather sorry that I cannot bring in all the varieties of non-conformers, as they have such interesting names—the red-giants, the white-dwarfs, the blue dwarfs, the black dwarfs, the sub-giants, the sub-dwarfs, and the collapsed super-giants.

SIZE AND DESTINY OF THE RED-GIANTS

The red-giants are normal enough so far as the amounts of energy radiated from their surfaces are concerned. Where they differ from main sequence stars is in being of much greater size, though of very much smaller average density. They really are big. It was these stars that I had in mind last week when I said that if the Sun were replaced by one of them it would fill up most of the space inside the Earth's orbit and that the Earth might even find itself lying deep inside the gigantic body of the star.

The line of argument I have described, starting from Eddington's calculations and ending with Bethe's discovery of the carbon-nitrogen reactions, fails completely for the red-giants. The reason is very simple. If Eddington's calculation is applied to them, it gives central temperatures that are far too low for the carbon-nitrogen reactions to work efficiently. Gamow suggested that perhaps other nuclear transformations are operative in these stars, but the difficulties raised by this hypothesis have proved so severe that it now appears to have been abandoned even by its author.

I have spoken of a different procedure from Eddington's, whereby Bethe's carbon-nitrogen reactions were put into the theory at the beginning and in which the size of the star was an outcome of the calculations. We were naturally interested to see whether stars as huge as the red-giants could be represented by the theory. We soon realized that this was impossible so long as the chemical composition was taken to be uniform throughout the star, as it was in Eddington's theory. Now the idea that every star has a uniform composition had been accepted by almost every author from the earliest attempts on the problems of stellar structure. Was there any reason for believing it to be a correct hypothesis?

In every normal star, hydrogen is being converted into helium inside the central regions, so that appreciable non-uniformity of composition must arise in any star that has already consumed a large fraction of its original supply of hydrogen, unless some process mixes up the helium with the material in the rest of the star. Given certain things in a star, a sufficiently fast rotation, and perhaps a certain sort of magnetic field, such uniform mixing does go on. When adequate mixing occurs the star remains of main sequence type; that is, its radius remains comparatively small. But when there is only partial mixing, it transpires that the star must swell as the hydrogen is consumed. Indeed, when 80 or 90 per cent of the inner hydrogen has been converted to helium, calculation shows that the distension of the star becomes exactly of the order observed in the red-giants. The importance of all this to the astrophysicist is that when we observe a star with a greatly distended bulk, we know that this star has had sufficient time since it was born to consume most of its initial supply of hydrogen. As we shall see next week, this result enables us to work out the ages of the stars with considerable accuracy.

In small stars such as the Sun, that have not yet lived

long enough to burn up much of their hydrogen, there can be no great degree of swelling. Even so, it seems that there has already been a slight expansion of the Sun, and this gives us the probable answer to our final astrophysical question. As the Sun steadily grills the Earth it will swell, at first slowly and then with increasing rapidity, until it swallows the inner planets one by one: first Mercury, then Venus, then the Earth. Mars is likely to be the last planet to suffer this fate, but it is just possible that an even further extension, as far as Jupiter, will occur. This particular part of the New Cosmology seems to fit in well with medieval ideas about hell.

ATOMIC ENERGY

My final points to-night are about terrestrial sources of energy and the possibility of blowing up the Earth. As to sources of energy, I think that it is now the popular idea that we should generate atomic energy on a large scale by the disintegration of uranium or thorium in a multitude of atomic piles, as they are usually called. Now in this matter I am heterodox in my views. I believe that the Sun is a far better bet, that we should trap sunlight on a large scale by the widespread growth of suitable plants in tropical areas. The ultimate product would be alcohol, an excellent substitute for petrol.

I am not alone in this belief, and I am not the first to express it. For instance, the case for it was put very forcibly about two years ago by Professor Pryce of Oxford. The idea that is now generally held is that so little uranium is required that the source of supply hardly needs to be considered. But this is not so. The main snag in using uranium or thorium is that rich ores of these metals are so extremely rare that they could only provide us with industrial power for a few centuries. It is true, as I said last week, that there is plenty of very thinly distributed uranium. But in order to use it, about as much rock would have to be mined each year as at present we mine coal. Then the rock would have to be crushed and the uranium extracted chemically. All this could no doubt be done, but I think it would require at least as much effort as the growing of plants for the production of alcohol. This is not to say that atomic energy may not be used for specialized purposes; for example, in transport. But these applications will, for the most part, turn on the very difficult problem of developing what physicists call fast reactors. These are a sort of cross between the violence of the atomic bomb and the slow production energy in a pile.

I come now to our final question. Is it possible to produce an atomic explosion that starts a chain reaction in the Earth itself? In particular, could some reaction fire off the hydrogen that is present in water, especially in the water of the oceans? If all the hydrogen in the oceans were suddenly converted into helium, the Earth would be vaporized practically instantaneously. The blaze of radiation produced would be as large as the total emission from the Sun added up throughout a whole year, and if there is life on Mars, it would rapidly be reduced to ashes. If you ever mention the end of the world, that is the sort of end you should have in mind.

A high temperature is necessary before hydrogen is affected by nuclear reactions. The highest temperature that can be produced on the Earth occurs in a volume a few centimetres across for a time of about a ten-millionth of a second during the explosion of a uranium bomb. This temperature is about 150,000,000° C., which is more than seven times greater than the temperature at the solar centre. The issue is whether a uranium bomb exploded under water would act as a detonator to the hydrogen in the water. In the autumn of 1945 I looked into this matter and decided that the high temperature produced by the bomb lasts for too short a time for this to be possible. This conclusion was later confirmed by the atom bomb trials at Bikini.

THE HYDROGEN BOMB

But before we leave this subject we must also consider the possibility of the underwater explosion of bombs more violent than the uranium bomb. These considerations have particular relevance to the possibility of making a hydrogen bomb. The idea of a hydrogen bomb is to produce an extremely rapid conversion of hydrogen into helium; to do what the Sun does, but to do it quickly. To do this, two conditions are necessary. One is a high temperature, and this could best be achieved by using a uranium bomb as a detonator. The other necessity is to find a far faster reaction than the main processes that occur in the Sun and the stars. At first sight it looks as though this is an impossibility, because any process that can be used on the Earth can also occur in the Sun. But this overlooks a crucial point. The fastest reacting substances are so extremely rare in the material of the stars that they are not important in astrophysics. On the Earth, however, these substances can be prepared artificially in the laboratory and in the atomic pile.

The most powerful reaction possible was discussed by Bethe in 1939. To understand this reaction, which must be the basis of the hydrogen bomb—if indeed a hydrogen bomb can be made—it is necessary to consider the isotopes of hydrogen. Again I hope that the term isotopes and the sort of argument I am going to use have become familiar from all the discussion of atomic energy that has gone on since 1945. Ordinary hydrogen, the form of hydrogen I have been considering so far, has a nucleus consisting of one particle, the proton. The second isotope of hydrogen, often called heavy hydrogen, has a nucleus that consists of two particles, a proton and a neutron. Some newspaper reports have suggested that this heavy hydrogen is the main constituent of the hydrogen bomb. This I do not believe. The important substance is the third isotope of hydrogen, usually called triton or tritium, which, as its name implies, has a nucleus containing three particles, a proton and two neutrons. The relevant reaction is that a proton can combine with triton to form a helium nucleus or alpha-particle.

There is a difficulty for a member of the general public like myself to work out whether a hydrogen bomb can be made, because one item of relevant information concerning the proton-tritium reaction has not been published: what the physicists call the radiative width for the reaction. But it is easy to make a shrewd guess. For, short of a fluke, the missing information for the proton-tritium reaction can be deduced from several very similar reactions for which accurate experimental data is available. These reactions are all characterized by the formation of alpha-particles. For example, the nucleus lithium 7

D

and a proton give two alpha-particles, and the nucleus boron 11 and a proton give three alpha-particles. By proceeding in this way, what seems to me to be a reliable calculation can be carried out. It shows that a tritium-proton mixture could indeed be exploded with tremendous violence by a uranium detonator.

One of the reactions I have just mentioned—namely, the formation of two alpha-particles from lithium and a proton—has been suggested in a number of press reports as being the basic reaction of the hydrogen bomb. I think that this idea is quite impossible, because a lithium-proton reaction could never go fast enough at the temperature produced by the explosion of uranium.

But to get back to the oceans, would the hydrogen bomb explode them? This answer I think is quite definitely, no. The importance of the hydrogen bomb from a military point of view is that it can be made as large as practical questions allow, whereas the uranium bomb is severely limited in size. So a hydrogen bomb of extremely great explosive power can be made if the necessary quantity of triton can be manufactured. But it is not the total amount of energy released by the bomb that decides whether the oceans will explode. The crucial quantity is the temperature produced, and curiously enough this must be nearly the same in the hydrogen bomb as in the uranium bomb. So we may conclude that although mankind may engage in foolish personal destruction, the Earth itself is safe.

There is a further important detail of the hydrogen bomb. Can efficient use be made of the gamma-rays produced by the triton-proton reaction? If these gamma-rays could be employed in causing the emission of neutrons and fast protons the detonation problem would be greatly simplified. Whether this is possible or not I just do not know, but if I felt that the possession of the hydrogen bomb was necessary for my defence this is an issue that I would look into very carefully.

A matter that I should like to discuss if time were available is whether anything is to be gained by trying to keep these questions secret. The futility of the attempts that have been made to do so is shown by the fact that at any time during the last few years any competent physicist, whatever his nationality, could have worked out everything I have told you in a few hours' calculation.

III

THE ORIGIN AND EVOLUTION
OF THE STARS

SEVERAL scenes in nature are of overpowering splendour. Sunrise or sunset, especially when seen in the high mountains, is one of them. So also is the sight of the stars in the heavens. The stars are best seen as a spectacle, not from everyday surroundings where trees and buildings, to say nothing of street lighting, distract the attention too much, but from a steep mountain side on a clear night, or from a ship at sea. Then the vault of heaven appears incredibly large and seems to be covered by an uncountable number of fiery points of light.

Surprisingly, the number of stars that actually can be seen at any time with the unaided eye is only a little over two thousand. These stars all belong to what is usually called the Galaxy, and it is about the Galaxy, our Galaxy, that I am now going to talk. The number of stars that can be seen increases very rapidly when you do not have to depend on the naked eye. With even a small telescope you can distinguish about a million stars; with large telescopes, like the ones at Mount Wilson, the number rises to well over a hundred million.

A glance at the sky will show you that the stars are not uniformly distributed over it. There is a bright band of light, that people call the Milky Way, running roughly overhead in which particularly large numbers are concentrated. The stars take on this appearance because the Galaxy is shaped like a disk. When you look at the Milky Way, you are looking along the disk, and you see a large number of distant stars. But when you look at other parts of the sky, you are looking out of the disk, and you then see only a comparatively few stars—these are just the ones that happen to lie close to us. It is because of their nearness that so many of these stars appear bright.

Now I should like to give you some idea of the size of the Milky Way, and of the distances between the stars. Ordinary units, such as the mile, are not much good for this purpose. As you know, in many astronomical discussions it is best to use light as a measure of distance. For instance, it takes light rather more than a second to travel from the Moon to the Earth, and we can speak of the distance of the Moon as being rather more than one light second. It takes light about eight minutes to travel to us from the Sun, and we say that the distance of the Sun is about eight light minutes. I think you will agree that it gives an extremely graphic description of the distances of the nearest stars when I say that light takes about three years to travel to us from them. And when you look at the Milky Way with a small telescope you can see to a distance of more than a thousand light years.

Here we reach the main turning point in this series of talks. So far I have been describing an essentially static picture of the planets, the stars, and the Galaxy. We have simply been looking at the astronomical scene as observation reveals it to us. But there is a whole lot of questions of a dynamic character that cannot be decided at all by observation. Where did the Galaxy come from? How are stars born within it? How were our Earth and the planets formed? What is going to be the ultimate fate of the stars? These questions are samples of the issues that I shall be considering from now on. We shall deal here with the origin of the stars, next week with the origin of the planets, and in the final talk with the origin of the Galaxy and of the Universe itself.

To make a beginning then, imagine yourself to be looking out across space at the stars of the Milky Way. Perhaps the most important feature of the New Cosmology is the realization that this space is not empty at all. Throughout the Milky Way there is a diffuse gas, usually called the interstellar gas. A gas, you will remember, is a swarm of separate atoms and simple molecules. By far the commonest element in the interstellar gas is hydrogen. Hydrogen atoms are more than a thousand times as numerous as all other atoms and molecules put together. As we shall come increasingly to understand, hydrogen is the basic material out of which the Universe is built.

Although I am only just starting on my main theme, I cannot resist a digression here. Even though elements like iron make up only a very small proportion of the interstellar gas they nevertheless have several interesting effects. One of these has an indirect impact on our cosmology. Tiny particles of dust condense out of the iron atoms in the interstellar gas, rather in the way water drops form in the clouds of the terrestrial atmosphere. These dust particles are a great nuisance to the observational astronomer, for they produce a sort of fog that limits his vision whenever he tries to look deep into the Milky Way. Thirty years ago it was thought that when we look out at the Milky Way we see the whole of the Galaxy. But we know now that this view is hopelessly wrong. The fog that I have just mentioned cuts down our vision so much that, instead of our being able to see the whole of the Galaxy, we only see about a hundredth part of it. To describe the way in which the effect of the fog has been investigated would take us too far off our main track. So I shall simply quote the results that have been reached, notably by the Dutch astronomer Oort and by Lindblad of Sweden.

It turns out that the Galaxy is a thin disk with a diameter of about 60,000 light years. This disk consists of stars—and gas. Near its centre the disk is very likely much thicker than it is at its edges, where it trails away very gradually. Whereabouts in the Galaxy do the Sun and the planets lie? The answer to this is near the edge of the disk. If you want to look towards the centre of the Galaxy you should seek out the great star clouds that lie in the constellation of Sagittarius, the Archer. But you will not see the centre, it is forever hidden from us by the fog we have just discussed. That is to say, you will not see it optically. As you may have heard, certain stars are powerful transmitters of radio waves. Radio waves can easily penetrate the fog, whereas light cannot. So if you really want to detect the centre of the Galaxy you should use radio waves and not light. This can be done.

But it is now high time that we got back to the interstellar gas. It is extremely rarefied. On an average over the whole Galaxy a match-box full of it would contain only about 100,000 atoms. This may be compared with the material in a star, like the Sun, where on the average a match-box full would contain about a million million million million atoms. Yet, in spite of this enormous difference of density, the total quantity of material comprising the whole interstellar gas seems to be appreciably greater than the material in all the stars put together. The reason for this surprising result is that the interstellar gas occupies a truly vast volume. The crucial consequence to be drawn from this new cosmological development is that it is the interstellar gas, not the stars, that rules the Galaxy. It controls the motions of the stars. It also controls their birth and the way in which they are allowed to grow.

THE BIRTH OF STARS

Now I must introduce you to the idea that this immense disk of gas and stars is in motion, that it is turning round in space like a great wheel. How then do the stars move? The main motion of a star consists of an orbit that is roughly a circle with its centre at the centre of the Galaxy. The Sun and the planets move together as a group around such an orbit. The speed of this motion is nearly 1,000,000 miles an hour. But in spite of this seemingly tremendous speed it nevertheless takes the Sun and its retinue of planets about 200,000,000 years to make a round trip of the Galaxy. At this stage I should like you to reflect on how many ways you are now moving through space. You have a speed of about 1,000 miles an hour round the polar axis of the Earth. You are rushing with the Earth at about 70,000 miles an hour along its pathway round the Sun. There are also some slight wobbles due to the gravitational attraction of the Moon and the other planets. On top of all this you have the huge speed of nearly 1,000,000 miles an hour due to your motion around the Galaxy.

There is an obvious analogy between the motion of the Earth around the Sun, and the motion of the Sun and planets around the Galaxy. But the analogy must not be pushed too far. In the solar system the Sun contains most of the material, and it lies at the centre, whereas there is no specially large blob at the centre of the Galaxy. Also in the solar system the planets move in nearly circular orbits of very different sizes, whereas in the Galaxy many stars have orbits of nearly the same size.

I said a moment ago that the interstellar gas controls the births of the stars. It is now our business to see how this happens. Astronomers are generally agreed that the Galaxy started its life as a rotating flat disk of gas with no stars in it. The gas was probably distributed much as it is now. That is to say, the diameter of the disk was about 60,000 light years, but its thickness was only about ten light years. A fairly good idea of the rotation of the disk can be got by thinking of the rotation of a thin wheel. Once again this analogy must not be followed too strictly. There would everywhere be small perturbations in the detailed motions of the various bits of gas, especially near the edge of the disk. To assume a complete absence of such perturbations would be rather like supposing that the flow of water in a whirlpool is entirely smooth, being devoid of ripples and small eddies.

How does a rotating disk of very diffuse gas give birth to compact stars? Such a disk would be what mathematicians call gravitationally unstable. That is to say, the attractive force of gravitation would exaggerate any irregularities that were present in it at the beginning. From this it can be deduced that the gas would have to break up into a large number of separate irregular clouds. This prediction, first made by Jeans, has been confirmed by observation, which shows that the interstellar gas is indeed composed of clouds. The distance across an individual cloud usually lies between ten and a hundred light years. That is to say, the clouds are small compared with the diameter of the disk, but not with its thickness. Once clouds have condensed like this, gravitation again exaggerates all the small initial irregularities that they happen to contain. So further condensation would take place in each cloud. At this point it is only necessary to say 'and so on,' for by repeating the condensation process a sufficient number of times, we must eventually arrive at the particularly dense sort of condensation that we call a star. To sum up the stages—first a whirling disk of gas, then eddies, clouds, condensations, and finally stars.

Granted, then, that gravitation must lead to the condensation of stars within the rotating galactic disk of gas, let us consider the simplest case of this happening. This is when a star is formed out of a roughly spherical blob of gas. On account of the very diffuse nature of the gas, it is clear that such a blob has to be enormously compressed before a star can be formed out of it. In fact, the blob has to condense to about a millionth of its original diameter. So compared with the gas clouds a star is a body of very small dimensions, and this explains why direct collisions or even close approaches between stars are events of extreme rarity, whereas collisions between the gas clouds themselves are quite common. In spite of the large numbers of the stars there is plenty of room for them to move about without seriously interfering with each other.

Why does a stellar condensation ever stop contracting? Perhaps I had better clear up this question before we go any further. As a condensation shrinks, its internal temperature rises, and when this becomes sufficiently high, energy begins to be generated in the interior. This is due to hydrogen being converted into helium by nuclear transmutations, as I explained last week. A stage is eventually reached when the energy so generated is adequate to balance the radiation escaping from the surface of the star. Contraction then ceases and the body becomes a normal star like the Sun.

It would be possible to stop at this point and to say that we have explained the origin of the stars. But there are several other questions that trouble the astrophysicist. For instance, we could ask: 'Why is it that all the interstellar gas has not yet been condensed into stars?' Or again: 'Why do stars possess some degree of rotation?' The answer to this question is connected with the fact that every cloud of gas and every stellar condensation is in orbital motion about the centre of the Galaxy. I cannot explain here exactly why it should be so, but it can be proved that this motion has the effect of generating a rotation in every condensation as it contracts. As I shall describe next week, this apparently innocent detail of the condensation process has a profound influence on the evolution of exploding stars.

Then there is the question: 'Does the condensation of a star cease once a compact stellar body has been formed, or does condensation continue indefinitely?' As I had something to do with answering this question myself, I should like to digress. One afternoon, just about eleven years ago, I was invited out to tea by Lyttleton, now one of my colleagues at Cambridge, whom I had then heard of only by name. It transpired that we were both interested in the problem of explaining the marked changes that have occurred in the climatic history of the Earth, about which I shall be saying more later. It further turned out that we had quite independently been thinking along very similar lines, and in the course of the discussion, by a sheer stroke of good fortune, we hit on the clue to the answer to my last question. Of course, scientific research is not only a matter of drinking tea. While indolence is very important in itself, some hard work has to go with it. But for myself I find it very difficult to work to set hours, or to work in a large team of scientists, as appears to be common in America, or to have to tailor my views to fit an official doctrine, as seems necessary in Russia.

Tunnelling through Gas

To get back to our question: What happens to a star once a compact stellar body has been formed? Owing to the ripples and eddies that are constantly present in the interstellar gas the star soon finds itself moving through the gas. But any such relative motion between star and gas is small compared with their common motion around the Galaxy. Instead of the star rushing through the gas, it drifts through. The situation is somewhat analogous to two people passing each other in the corridor of a fast-moving train. There is a relative motion between them, but this is small compared with the speed of the train.

As it drifts like this through the interstellar gas a star tends to pick up more of it. The rate at which it attracts the gas and the way it picks it up can be studied mathematically. It turns out that the gravitational field of the star pulls in gas from far and wide, and as the star moves through the gas it leaves a huge empty tunnel behind it. The distance across the tunnel is enormously greater than the size of the star. The exact value of the diameter of the tunnel depends on the speed of the star through the gas. The smaller the speed the broader the tunnel. If you would like to understand this in a little more detail, imagine yourself to be seated on the star, and imagine yourself watching the interstellar gas as it streams past you. Then it is clear that the slower the gas moves the easier it will be for the gravitational pull of the star to overcome the energy of motion of the gas.

Perhaps this is the first time you have heard of these tunnels, but I can assure you that they are very important in the New Cosmology, and from now on I shall be saying a great deal about them. The tunnelling process evidently increases the quantity of gas within the star. How long does the star continue to grow? The answer is that tunnelling cannot stop so long as the star is immersed in gas. But a more important question is this: Can the quantity of material in a star be increased to a really marked extent in this way? Well, the star will not increase much unless its speed through the gas is exceptionally small, and by small I mean not more than 5,000 miles an hour. When the relative speed of star and gas is about 30,000 miles an hour, as it is for most stars, the tunnel is too thin for the changes to be appreciable, even over long time intervals. Notice, by the way, that a speed of 30,000 miles an hour is still small compared with the common orbital speed of about 1,000,000 miles an hour around the Galaxy.

Only about one star in a million is more than ten times as massive as the Sun. Lyttleton and I believe that these stars are just the ones that have had particularly low speeds through the gas during the last 100,000,000 years. Accordingly these stars have been drilling out extremely fat tunnels and have swept up tremendous quantities of interstellar gas, and that is why they are big. You will probably wish to ask whether our Sun is tunnelling out interstellar gas at the present time. Lyttleton, Bondi, and I think that the Sun certainly is sweeping up gas. I will describe one of the reasons why we feel pretty confident about this. If you look at the Sun under normal conditions—for instance, at sunset—it appears to have a sharply defined surface. This is the surface that radiates most of the light and heat into surrounding space. It is this surface that, as I said last week, is at a temperature of about 6,000° C. Now during a total eclipse of the Sun you also see a faint delicate outer atmosphere, an atmosphere with two parts. The inner part is known as the chromosphere and the much more extensive outer part is called the corona. You may have seen photographs—and very striking photographs they are—taken during an eclipse of the Sun.

In the total solar eclipse of 1878 streamers in the corona were observed to stretch as far as 5,000,000 miles from the Sun. These streamers were simply captured interstellar gas falling into the Sun. They had the appearance of gigantic flames because they were made visible by the power of the Sun's rays. It must not be thought, however, that the whole of the corona is composed of material that falls into the Sun in this way. For the captured interstellar gas is moving with very high speed, and as it collides with the outer parts of the Sun it produces a gigantic splash everywhere on the Sun's surface. The distinction between the infalling material itself and the splash is clearly visible to the astronomer, who refers to the splash as the inner corona and to the infalling material as the outer corona. There is much of interest that could be said about the atmosphere and surface of the Sun, but such matters lie rather outside the topics that form the substance of these talks.

We can use observations of the solar corona to estimate the size of the tunnel drilled out by the Sun. It turns out that the diameter of the Sun's tunnel is at present about 1,000 times greater than the diameter of the Sun itself. Large as this may seem it is really rather a thin tunnel. It is certainly too thin for the amount of material in the Sun to increase appreciably, even if the process were to go on for as long as 10,000,000,000 years. The reason for this is that the Sun's present speed through the interstellar gas is much too high for a fat tunnel to be really possible. But the Sun's speed through the gas must be changing continuously, due to eddies and other disturbances within the gas. So this result applies only to the present. Was the Sun's tunnel fatter in the past? What is the chance of it becoming fatter in the future?

Let us take the past first. There is little doubt that at several periods in the Sun's history the tunnel must have been very much wider than it is now. This can be deduced from the climatological history of the Earth, which provides clear evidence that at certain times the Sun must have been considerably warmer than it is at present. For example, coal is found in Spitzbergen within 12° of the North Pole. Now this coal required the prolific growth of a type of plant normally associated with tropical conditions. Or again, fossilized plants have been found near the South Pole. It also seems possible that the famous problem of the cause of the Ice Ages can be solved along those lines. Curiously enough, meteorologists suggest that an increase, not a decrease, of the Sun's heat is needed to produce an Ice Age. The necessary increase is, of course, much less than would be required for plants to grow at the poles. The main points of the meteorologists are these: an increase in the Sun's heat would produce an increase in the cloudiness of the polar regions, and this leads to a more equable climate; that is to say, the winter temperature is raised and the summer temperature is lowered, provided the increase in the Sun's radiation is not too large.

The reason for the lowering of the summer temperature is that the clouds reflect an increased proportion of the Sun's light back into space and prevent it from ever reaching the ground. Now the crucial requirement for the formation of huge ice sheets is a lowering of the summer temperature. For in many places great quantities of snow are deposited during the winter, but no glaciers are formed because the snow melts away in a month or two near midsummer. A more equable climate in such places must lead to the formation of permanent ice sheets.

It is known that the Sun cannot have been warmer during these periods as a result of changes in its own

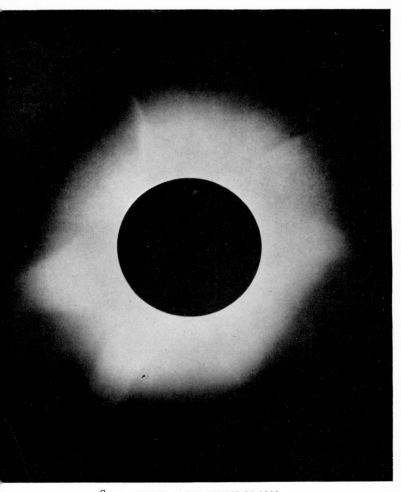

SOLAR CORONA IN THE ECLIPSE OF 1929

(*Mount Wilson and Palomar Observatories*)

internal structure. It seems likely that the excess radiation necessary to produce the climatic changes I have just described was due instead to the infall of interstellar gas on to the solar surface. I have explained that the Sun sweeps up all the gas lying inside a tunnel. As it approaches the Sun the speed of infall of this material increases and eventually it rains on to the solar surface at speeds of more than 1,000,000 miles an hour. As you know, when a moving body is stopped by impact the energy of its motion is converted into heat. Exactly the same thing happens when the gas falls into the Sun. Its impact with the Sun produces heat, and the effect of this is an augmentation of the Sun's normal emission of radiation into surrounding space. That is to say, the Sun emits more radiation than it would do if it were not tunnelling out interstellar gas.

At present the augmentation does not have very much effect on the light and heat reaching the surface of the Earth. But even though the tunnelling process is weak at the present time it does have one important effect. Without it the whole problem of broadcasting would be very much harder. The reason for this is that the extra radiation arising from the gas falling into the Sun consists in the main of ultra-violet light and mild X-rays, and these are responsible for producing what are called the ionized layers in the Earth's atmosphere. Without these layers broadcasting would be more difficult and far more limited than it is. So we owe a very practical debt to the tunnelling of interstellar gas by the Sun.

To return to the climatological history of the Earth: to produce a climate hot enough for tropical plants to

E

grow near the poles the diameter of the tunnel drilled by the Sun would have to be more than a hundred times wider than it is at present. For this to occur the Sun's speed through the interstellar gas must have at one time been no more than about 5,000 miles an hour. All this happened, according to geological evidence, about 200,000,000 years ago. It is as well for us that the tunnel did not stay as fat as it was then, otherwise the Sun's mass would have increased appreciably during the last 200,000,000 years, and by now the Sun would have got too warm for our comfort. Luckily this did not happen, but the occurrence of four Ice Ages during the last million years shows that the width of the tunnel has been varying quite a lot lately. As I have just said, the tunnel is quite small now, its width is only a thousand times the diameter of the Sun itself, but if the tunnel should widen out again, we shall be due for another Ice Age when the great northern glaciers will spread out and will cover once again the face of our country. If the tunnel should widen out still further, the Sun's heat will become so great that the ice sheets will melt and tropical conditions will spread to the poles of the Earth.

Last week I said that if the Sun is not much changed by processes outside itself, it will remain much as it is at present for the next 10,000,000,000 years. Then, because of internal changes, it is certain to grow gradually hotter and will grill the Earth until the oceans boil. What I had in mind when I talked about outside processes was that the Sun's tunnel might widen out owing to changes that are constantly taking place in the motion of the interstellar gas. If the tunnel should widen out to a really large size and should stay like that for the next 100,000,000 years, the amount of gas swept up will appreciably increase the Sun's mass. This is not very likely to happen, but it is of interest to consider its consequences if it does. There is a chance of about one in a hundred that the Sun will increase its mass threefold or fourfold in the next 1,000,000,000 years. Such an increase of mass would lead to about a hundredfold increase in brightness, and this would just about melt the rocks of which the Earth is composed. At any rate they would become sticky. There is a chance of about one in 10,000 that the mass of the Sun will increase about twentyfold from this cause. The consequent increase of brightness would then be about ten thousandfold and this would just about vapourize the Earth. The figures I have given you show the chance of the Sun entering on a spectacular career. They are small, but not negligibly small. In fact, they are considerably larger than the chance of winning a big sweepstake. But, whether this happens or not, the fate of life on the Earth will be the same. As I said last week, we shall certainly be roasted. These questions are only important in deciding whether this will occur sooner or later.

You will understand that what I have been able to say about the condensation of gas into stars only concerns the main issues. There are many other consequences of the ideas we have been discussing. I shall have no time to say anything about most of these other matters. I can just squeeze time to mention that the Galaxy has a large number of satellites. These satellites are not single bodies like the satellites in the solar system, but gigantic clusters that each contain more than 100,000 stars. They are usually referred to as globular clusters, a name derived from their spherical appearance. But, vast as the globular clusters are, they are genuine satellites in the sense that they are small compared with the Galaxy itself. At present nearly all of them lie outside the disk that forms the Galaxy. But they move in orbits because of the gravitational attraction of the Galaxy, and we can show that these orbits are of such shapes that the clusters must pass through the galactic disk from time to time. It is not impossible that one day a globular cluster might pass through the particular bit of the Galaxy in which we are located. If this should happen more than a thousand stars as bright as Sirius could be seen, and there would be a moderate chance that a star belonging to the cluster might come close enough to appear as bright as the full Moon.

There are about a hundred of these globular clusters and each of them must have passed through the galactic disk many times. As you will readily perceive, the effect of all this is to cause a good deal of localized disturbance within the Galaxy. To begin with the disturbances affect both stars and gas, but the gas recovers comparatively quickly and reverts substantially to its initial state, whereas the stars do not. I cannot explain what happens in precise detail, but the main effect on the Galaxy is that the disk of stars becomes thicker than the disk of gas. Points such as this are of interest to the astrophysicist, because it means that quite a high proportion of the stars in the Galaxy have suffered disturbances that have pushed them out of the interstellar gas, or at any rate have pushed them out of the gas for most of their lifetimes. These stars, there-fore, cannot grow much by the tunnelling process. This detail has importance, as I shall mention later, in our method of finding the ages of the stars.

Another interesting question that I cannot discuss in any detail is this: What happens if a star passes through one of the clouds of dust, which as I said earlier also occur within the disk of the Galaxy? This problem has been investigated by Lyttleton, and he finds that if the speed of the star through the dust cloud is sufficiently small it will capture great quantities of it, and form them into separate, loose bundles of particles. These loose bundles are the comets. It is more than probable that it was in this way that the thousands of comets which move round the Sun have come into being.

MULTIPLE STARS

There is one other consequence of all this tunnelling that is so important that we simply must give some attention to it. Without the process I am now going to mention none of you could be listening to me. Without it the Earth and the planets could not have been formed. We shall only be able to start the argument now; the main details will have to come next week. For the most part two neighbouring stars do not stay together throughout their lifetimes. But from time to time circumstances arise when this does happen. We then speak of the two stars as having formed a double system. To begin with the distance between the two components, as the two stars are usually called, is not very different from the normal spacing of neighbouring stars. But the components of a double system sweep out two roughly parallel tunnels in the interstellar gas, assuming of course that they happen to lie inside the gas, and the effect of this tunnelling is to bring the two stars closer together. At first the two tunnels are separate, but if the process continues long enough a stage is reached when the tunnels merge together, and the double system, or binary as it is often called, has simply one tunnel, the proceeds of which are shared by the two stars.

As the components come together they move in orbits around each other, orbits that are at first flattish ellipses but which gradually become circular in shape. About a thirtyfold increase in the mass of the system is sufficient to take the two stars from an initial separation of, say, a tenth of a light year down to a separation of only a few light minutes, or even less than that. The time required for the two components to go once around each other also changes from an initial value that may perhaps be as great as 100,000 years down to a period of only a day or two. When this stage is reached the two stars are so near each other that they are practically in contact. It is natural to ask whether this does not continue until the two components actually merge together into one star. I think this does happen, but what the fate of such an amalgamated star is likely to be is just another of those topics that lie outside the scope of these talks.

Let us turn now to observation. Observation shows that binary systems are extremely numerous. In fact, about as many stars belong to double systems as there are single stars like the Sun. This gives you a measure of the importance and universal application of the tunnelling process. The various double systems observed show all stages of the evolutionary sequence I have just discussed. That is to say, the separation of the two components varies in the different systems from extremely large values down to cases where the two stars practically touch each other. Multiple systems containing more than two components also arise. For instance, a binary can join with a third star to form a triple system. Quadruple systems can be formed in two ways, either by joining two binaries or by a fourth star joining with a triple system. Both these types are observed. Still more complicated systems can be formed. The prize specimen is Castor, of the pair Castor and Pollux, which contains six stellar components. And the famous Pole Star, instead of being one star as it appears to be, is really five. Even larger groups can be formed. The same tunnelling process also explains the formation of some of the groups of stars that you see in the sky. For instance, it is probable that the well-known group, the Pleiades, were produced in this way. Since the main stars of the Pleiades are large, we can infer that the tunnelling of interstellar gas has been very important in this cluster. In fact, it is probably in full swing at the present time.

ESTIMATING THE AGES OF THE STARS

Throughout this talk we have been concerned with the passage of time; in other words, with the evolution of the stars. How long have these processes been going on? The astrophysicist grapples with this question by considering the nuclear processes that lead to hydrogen being converted into helium inside the stars. By a fairly straightforward calculation he can find the time required for these nuclear processes to consume a given quantity of hydrogen. For instance, in a star as massive as the Sun the hydrogen supply would last for about 50,000,000,000 years. Since nearly all the solar hydrogen is still unconsumed we believe that the Sun cannot be anything like as old as this.

The main danger we have to avoid is in getting our age estimates confused by the fact that a star can constantly refuel itself by sweeping up further supplies of interstellar hydrogen. To avoid this as far as possible, we consider only stars that now move for the most parts in regions where there is little or no gas. And we can further restrict the choice to stars that even during the brief periods when they do move through gas of appreciable density do so at comparatively high speeds. We next look for stars that have just about reached the end of their store of hydrogen. As I explained last week, we can recognize stars in this state; they are known as the red-giants. These stars are particularly large in size, and we know that a star cannot have a very large volume unless most of its hydrogen has been used up. It then only remains to estimate the amount of material present in each star, and this can be done with considerable accuracy by using their observed brightnesses. So finally an estimate of the age of each red-giant can be obtained. This has been carried out for a large number of cases. The results are very satisfactory. No estimate exceeds about 4,000,000,000 years, although a lot of stars do come close to this value. So it seems that 4,000,000,000 years is a pretty good value for the ages of the oldest stars. This means that even if the Sun belongs to the oldest stars it has still only lived long enough for about twenty trips around the Galaxy. So you will realize that the Galaxy is still very much in its early youth. If instead of our only getting a peep at it, we could observe the Galaxy throughout two or three revolutions of the Sun in its orbit, we should see that, so far from being a worn-out structure, the Galaxy is really a scene of violent and never ceasing activity. We should, in fact, see the dynamic picture that the New Cosmology presents to us.

For comparison with all this I might mention the estimates that geophysicists give for the age of the Earth. I cannot enter into any detail now, but this can be done by a variety of methods most of which depend on the radioactivity of the uranium present in the rocks of the Earth's crust. There is some disagreement about the results, but this is not important for our purpose. It is sufficient to say that almost all the estimates come out between 2,000,000,000 years and 3,500,000,000 years. All these results when taken together are good enough for us to say that the Earth, and presumably the other planets, are younger, but not very much younger, than the stars.

When I was at school I learnt history in such a way as to think a period of a century or two was a very long time. It came as a great shock to realize later that the real history of man must be measured not in centuries but in tens and perhaps in hundreds of thousands of years. But even this is only the briefest tick of the clock compared with the ages of the rocks in your garden and the stars in the sky. What is so important about the time estimates of the astrophysicist is not that the results are staggering almost beyond belief, but that they are quite definite and precise, more precise than anything we know about the history of man if you go back more than a few thousand years. We are quite definitely faced with the situation that our Galaxy is not a timeless structure, but something that came into being about 5,000,000,000 years ago. How did it come into being? What is the significance of periods of time like this? These are the deeper issues that come out of our present discussion. The answers to them must form a part of our cosmology when later we come to consider the Universe as a whole.

IV

ORIGIN OF THE EARTH AND THE PLANETS

IN my first talk I referred to observations by Lyot concerning the clouds of Venus. It has been pointed out to me that the suggestion that the clouds of Venus might well be colossal swarms of dust particles was put forward by Mr. J. Evershed as long ago as 1918. As Mr. Evershed's remarks seem to have been quite generally overlooked in present-day astronomical literature, I am very glad to have this opportunity to refer to his work.

TORN OUT OF THE SUN?

Now to begin our main discussion. My overall plan in these talks has been to spread outwards from ourselves. First we considered the Earth and the other planets of the solar system, then we moved on to the Sun and the stars, and last week we saw that the Milky Way is but a small portion of a huge disk-shaped cloud of gas and stars that is turning in space like a great wheel. The next stage, if we were to follow our plan rigidly, would be to step outside the Galaxy, as this great disk is called, and to see what further bodies are to be found in the depths of space. But before we come finally to these things I am sure you will expect me to deal with how the Earth and the planets came into being. Perhaps you will find it surprising to hear that this matter is closely related to the ultimate fate of the stars. This also is an important issue that we must now consider. The origin of the planets is one of the high points of the New Cosmology. It affects our whole outlook on life. For instance, the question of whether life is rare or commonplace in the Universe depends essentially on this issue. I suppose that it is because of its cosmological importance that many people are given so strongly to asserting that the planets originated as bits of material that were torn out of the Sun. For some reason or other this idea has a deep-rooted appeal. So perhaps I had better begin by outlining some of the arguments that show why it must be wrong.

An appreciation of the scale of the solar system is very important if we are to understand its origin. As I have said before, this can best be done by making a model in which the Sun is represented by a ball about the size of a large grapefruit. On this model the great bulk of the planetary material lies at a hundred yards or more from the Sun. In other words, nearly all the planetary material lies very far out. This simple fact is already the death blow of every theory that seeks for an origin of the planets in the Sun itself. For how could the material have been flung out so far? As an instance of this difficulty, H. N. Russell found that if Jeans' well-known tidal theory were right, the planets would have to move around the Sun at distances on our model of not more than a few feet. This notion of Jeans', which still seems to be very widely believed, was that the planets were torn out of the Sun by the gravitational pull of a star that passed close by.

Once this difficulty was appreciated, people attached to the planets-from-the-Sun idea shifted their ground.

The planets, they said, were not formed with the Sun in a state like it is at present, but at a time when the Sun had a vastly greater size, as it must have had at the time when it was born. But it is hard to see how anything can be achieved by this modification. To make it work at all it would be necessary to demonstrate that a blob of primeval gas—the interstellar gas you may remember from last week—could condense in such a way that the great bulk of it went to form a massive inner body—that is to say, the Sun—surrounded at vast distances by a wisp of planetary material. And I do not think that this can be done. At any rate all the attempts that have so far been made to cope with the difficulty fall very far short of the mark. But instead of stressing this, I should like to give you another and perhaps more important reason why our Earth and the planets cannot have originated with the Sun.

I have tried to bring out in these talks the dominating cosmic role played by hydrogen, the simplest of the elements. Helium, the next simplest, is produced in appreciable quantities in the inner regions of normal stars like the Sun. But, apart from hydrogen and helium, all other elements are extremely rare, amounting as they do to no more than a hundredth part of the mass of the Sun. Contrast this with the Earth and the other planets, where the opposite situation occurs, where hydrogen and helium make a smaller contribution than highly complex atoms like iron, calcium, silicon, magnesium, and aluminium. This contrast brings out two important points. First, we see that material torn from the Sun would not be at all suitable for the formation of the planets as we know them. Its composition would be hopelessly wrong. And our second point in this contrast is that it is the Sun that is normal and the Earth that is the freak. The interstellar gas and most of the stars are composed of material like the Sun, not like the Earth. You must understand that, cosmically speaking, the room you are now sitting in is made of the wrong stuff. You, yourself, are odd. You are a rarity, a cosmic collector's piece.

Here then is a way to approach the problem of the origin of the planets. We must find a source of the strangely complicated rare material out of which the Earth and the planets are made. Let me first outline very briefly the solution of the problem, and then we can look back into details. To put it in two sentences, there was once a star moving around the Sun that disintegrated with extreme violence. So great was the explosion that all the remnants were blown a long way from the Sun into space with the exception of a tiny wisp of gas out of which our planets have condensed.

If you pick a star at random the chance that it will be a separate star by itself, like the Sun is at present, is no greater than the chance that it will be a member of a binary system. A binary system, you will remember, contains two stellar components that pursue orbits around each other. Let us see what can happen if we suppose that the Sun was at one time a component in such a double system. We must now make a choice for the distance between the Sun and the companion star it used to have. It is important to realize that there is practically no restriction on our freedom of choice here, because, as I explained last week, the distance apart of the component stars in a binary may be anywhere in the enormous range from a tenth of a light year down to a fraction of a light minute. The required distance apart of the Sun and its companion star is intermediate between these extremes, being about one light hour. That is to say, on a plan with the Sun represented by our grapefruit the companion star would be about 100 yards away. This value will give you a clue as to how the choice of separation is made; namely, so that in the final outcome the bigger planets will be found to lie at the right sort of distances from the Sun.

Explosion of Supernova

The next step is to draw up a set of specifications for the companion star. It must have been appreciably more massive than the Sun. It must have been a very special star. It must have been a star that exploded with extreme violence. Such stars are known to the observational astronomer as supernovae. Thanks largely to the work of the two Mount Wilson astronomers, Baade and Minkowski, we know a good deal about these explosions. Most of the material of a supernova—that is to say, considerably more material than there is inside the whole of the Sun—gets blown out into space as a tremendous cloud of fiercely incandescent gas moving at a speed of several million miles an hour. For a few days the accompanying blaze of light is as great as the total radiation by all the 10,000,000,000 or so stars in the Galaxy. It was out of such a holocaust that the Earth and planets were born, and it happened in this way.

Not all of a supernova is blown away as gas in such an explosion. But the dense stellar nucleus that was left over after the explosion of the Sun's companion star did not stay with the Sun. One of the effects of the explosion was to give this stellar nucleus a recoil that broke its gravitational connection with the Sun. It moved off and is now some unrecognized star lying in some distant part of the Galaxy. But before it left the Sun, and during the last dying stages of the explosion, it puffed out a cloud of gas that the Sun managed to hold on to. In as little as a few centuries this cloud of gas spread out around the Sun and took on the form of a rotating circular disk. As I shall be describing later, the planets condensed out of the material in this disk. So we see that the real parent of the Earth is not the Sun at all, but some unknown star that is probably unnamed and unseen.

According to the results of Baade and Minkowski the temperature inside a supernova is about 300 times greater than it is at the centre of the Sun. At such a temperature all manner of nuclear transmutations occur with great rapidity. The helium-hydrogen reactions which are so important in the Sun are no longer important here. Instead, helium becomes transmuted to elements of what is called high atomic weight; for example, magnesium, aluminium, silicon, iron, lead, and uranium, to name only a few. The importance of this is obvious. It means that the companion star's final gift to the Sun was a cloud of gas with just the right kind of composition necessary to account for the constitution of the Earth and the planets.

Before we go on to discuss the cause of a supernova explosion, perhaps I might mention how this general picture of the origin of the planets has arisen. It is really the outcome of developments that started with Jeans' tidal theory. First this was modified and improved by Jeffreys. Then H. N. Russell overthrew both these theories with the sort of criticism I referred to

5'

GALAXY M 81 IN URSA MAJOR

(*Mount Wilson and Palomar Observatories*

earlier. Lyttleton was the next to take up the problem about fifteen years ago. He was the first to realize for certain that the planetary material cannot have come out of the Sun, and it is to him that we owe the development of the double star idea. Once this stage was reached the remaining steps were more or less inevitable, and as it happens it was I who took them. They arose for the most part through an attempt to put the theory on a firm observational footing.

ROTATIONAL INSTABILITY

What was the cause of the supernova explosion? Why, in contrast with the spectacular career of its companion, has the Sun remained a normal well-behaved star? The answers to these questions lie implicitly in our requirement that the companion must have been appreciably more massive than the Sun. It is very likely that the companion grew to this great size through sweeping up large quantities of interstellar gas by means of the tunnelling process I described last week. The same sweeping up of gas has led to the Sun growing more massive than the average star and it was also this that was responsible for the formation of the binary system in the first place. But the sweeping up of interstellar gas is a chancey business. It occurs in fits and starts. Sometimes it is particularly strong and sometimes it is very feeble. It is quite possible for a star to grow extremely massive and then for the sweeping up process to be suddenly cut down by some vicissitude affecting the interstellar gas. This must have happened to the Sun's companion.

To begin with the companion was a normal star—that is to say, energy was produced in its interior

F

through hydrogen being converted into helium at just such a rate as to balance the radiation of the star into surrounding space. But to achieve this balance the companion had to consume its hydrogen at a very much faster rate than the Sun. To be precise, if the companion were ten times as massive as the Sun, the hydrogen would have to be used up a thousand times faster. So the hydrogen supply in the companion could only have lasted for about one-hundredth part of the corresponding period for the Sun. Calculation shows that the respective times are about 500,000,000 years for the companion and about 50,000,000,000 years for the Sun. Compare this with the present ages of the oldest stars in the Galaxy, which were born about 4,000,000,000 years ago, and you see that, whereas the Sun has not yet had time to consume much of its hydrogen, the companion star had ample time. Such massive brilliant stars that quickly squander their resources have a special name—they are called the supergiants.

Our next step is to consider what happened once all the hydrogen in the interior of the companion star had been used up. The leakage of energy from the centre to the surface of such a star does not in itself require energy to be produced by the conversion of hydrogen into helium in the inner regions. For this leakage depends only on the central temperature and on the chemical composition and amount of material within the star. So, although energy production ceased in the interior of the companion star, there could have been no corresponding interruption in the outward flow of energy from the central regions to the surface. This energy was ultimately radiated into space from the surface and was therefore entirely lost from the companion star. The loss had to be made good by the star slowly collapsing inwards on itself. In other words, the companion developed into a collapsed supergiant. As it did so the central temperature necessarily became greater, and the leakage to the surface also became greater. So the first effect of the loss of radiant energy at the surface was not to cool off the star but to heat it up. This was only achieved, however, through the star living on its capital, through it collapsing inwards on itself.

When did this contraction cease? To answer this question I must now remind you that every star is in rotation. And by a well-known principle in mechanics, as the companion star collapsed its rotation became more and more rapid. As it did so the internal forces set up by the rotation became larger and larger. This could not go on indefinitely. A stage has to be reached at which the rotary forces become comparable with gravity itself. At this stage such a star begins to break up through the power of its own rotation. But this is not the end of the story. We must look a little deeper into the contraction process if we are to understand the incredible violence of a supernova explosion.

So long as the radiation that escapes from the surface of a star like this is the sole cause of the collapse nothing very violent can happen. The rotary forces increase too slowly for that. What happens is that the star breaks up not in one enormous explosion, but through the steady showering off of material rather like a gigantic catherine wheel. The steadiness of this process is occasionally punctuated by a sort of spluttering in which a cloud of material, roughly comparable with the Earth in total mass, is ejected into space with a speed of about 10,000,000 miles an hour. When this happens the hot inner regions of a star that undergoes this process are temporarily exposed and this leads to a transient increase in its brightness. Such occurrences are familiar to the astronomer who refers to them as ordinary novae. But the explosion of a supernova is on a far grander scale than this.

We must now see how supernovae arise. I have mentioned that as a collapsed supergiant shrinks its internal temperature becomes greater. Reactions between atomic nuclei must become more rapid as the temperature rises. When the temperature has risen about a hundredfold the conversion of helium into heavy elements like iron must become very important. This is the process of the generation of the elements that I referred to earlier. Now, if the collapse proceeds far enough before the rotary forces break up such a star, these nuclear reactions must start to *absorb* energy instead of generating it. This situation, which now goes the opposite way from everything we have considered so far, is due to the large-scale production of free neutrons. When this stage is reached the loss of radiation from the surface becomes by comparison quite unimportant, and the star then collapses catastrophically because of a rapid absorption of energy by the nuclear processes and not because of a slow loss of energy at its surface. Instead of being slow and steady, taking hundreds of thousands of years, the collapse becomes swift and catastrophic. The rotary forces now grow rapidly until they become so large that the collapse of most of the star is halted and a large part of it gets flung out into space in a supernova explosion.

So the stages in the production of a supernova are these: first, a massive supergiant exhausts its supply of hydrogen, then the supergiant begins to collapse because of the continual loss of energy through radiation from its surface. As contraction proceeds, rotation becomes more important. The final requirement is that rotation must not break up the star until after the absorption of energy by nuclear reactions has brought about a catastrophic collapse. Otherwise the star will simply splutter its way through a long series of ordinary nova eruptions instead of reserving the whole break-up process for one colossal explosion.

So this is the sort of thing that happened to the parent of our planets. Calculation tells us a good deal about its state just before the outburst. The collapse must have gone on very far before this happened. In spite of the enormous amount of material in the companion star, it must have become considerably smaller in volume than the Earth. It emitted hard X-rays from its surface into surrounding space. It was so enormously dense that a match-box full of material taken from its central regions would have contained about 1,000,000,000 tons. Its surface rotated with a speed of about 100,000,000 miles an hour. And the time required for its catastrophic outburst was as little as one minute.

So much then for the parent of our planet. Supernovae have other interests for the astrophysicist. As a recurrent theme in these talks we have seen that hydrogen is the basic material out of which the universe is built. Helium is common in stars compared with other elements because it is produced in appreciable quantities inside them. The abundances of the rest of the elements are so small that it is natural to ask whether all the material in the Universe started its life as hydrogen. It seems to me very likely that this is correct. I think that the other atoms have all been produced within the stars, in particular that the heavy elements such as iron have been built up in the dense collapsed supergiants we have just been describing. The explosions of these stars distribute material in interstellar space, where some of it forms into the great clouds of dust particles that we discern with the telescope. It is also likely that some of the material escapes altogether from the Galaxy into surrounding space. I shall mention this again next week when we come to consider the origin of the Galaxy itself. Then there are other questions, such as whether the collapsed supergiants are powerful transmitters of radio waves, and whether the supernova explosions are the main source of the mysterious cosmic rays whose energies are so extraordinary. I shall have to pass these interesting questions with the brief remark that it seems very likely to me that the answer to them both is, yes.

THE CONDENSATION OF THE PLANETS

But now we must get back to the comparatively tranquil final stages in the formation of the planets that followed the explosion and that occupied a period of about 1,000,000,000 years. A few centuries after the outburst of the Sun's companion, the stellar remnant of the explosion must have travelled very far away from the Sun, or at least far enough for its effect on the wisp of gas that was captured by the Sun to be unimportant. This wisp of gas then settled down into a flat circular disk that rotated around the Sun—that is to say, it spread out around the Sun and then it settled down into the disk. The main part of the gas must have been distributed in the regions where the orbits of the great planets, Jupiter, Saturn, Uranus, and Neptune, now lie. This means that on the model we are using with a grapefruit sun the main part of the disk must have had a diameter of several hundred yards. At its edges the gas would have trailed away very gradually.

But I must now explain how such a rotating disk of gas condensed into the planets as we know them. Once the supernova remnant had receded to an appreciable distance, the temperature of the main bulk of the gas in the disk must have fallen well below the freezing point of water. Many sorts of molecules must then have been formed and, as was pointed out in 1944 by Professor Jeffreys and A. L. Parsons, these molecules must have collected into a swarm of solid bodies by a process very similar to the formation of water drops in the clouds of our own terrestrial atmosphere. This condensation into solid bodies must have been offset to some extent by collisions between the bodies themselves, which tend to return material to the gaseous state.

At any particular time there must have been a rough balance between condensation from gas into solid bodies and evaporation that converted solid material back into gas. You might think that this stalemate would have had to go on for ever, and it probably would have done if the raindrop form of condensation were the whole story. But if any condensation should ever grow to a certain critical size, which is about *a hundred miles across*, the gravitational pull of the condensation itself begins to play a dominating role. The gravitational field is, so to speak, able to reach out into the surrounding gas and drag it in on to the condensation. When this happens the rate of condensation is greatly increased. It is this that ensures that the gas will form into a few comparatively large bodies rather than into a swarm of much smaller particles. The essential point is that the chance of a particular body ever growing to the critical size is very small, but given sufficient time it will certainly happen in a few cases. The fewer the number of cases the fewer the number of planets into which the material finally condenses. For once the gravitational field of a growing body comes into operation the rate of acquisition of material becomes so large that the first few bodies to attain the critical size then go on to snatch up practically all the material of the disk.

Perhaps you will see this best if I quote one or two of the results calculated for our own solar system. The first condensations to grow large take about 1,000,000,000 years to reach the mass of the Earth. But from this stage only about 100,000 years was needed for a primordial planet to increase its mass up to the same order as those of the great planets Jupiter, Saturn, Uranus, and Neptune. This shows you the tremendous accelerating effect of the gravitational condensation. An important consequence of this is that the Earth can hardly have been formed as a primordial condensation, a result already inferred on other grounds by Lyttleton more than ten years ago. For a condensation would hardly stop short after taking 1,000,000,000 years to reach the mass of the Earth if it only needed a further 100,000 years to go on and become a great planet. It could, of course, be argued that a condensation stopped short at the mass of the Earth simply because all the gas in its neighbourhood had become exhausted. This might be a reasonable argument if we had only one case to explain, but there are five planets—Mercury, Venus, Mars, Pluto, and the Earth—together with about thirty satellites also to be accounted for. It would be stretching coincidence much too far to suggest that exhaustion of material was responsible for cessation of growth in all these cases.

Besides, there is another argument that shows the same thing. None of the present planets can have been primordial condensations, not even the great planets. For owing to the rotation of the disk around the Sun, the primordial condensations must have acquired axial rotations—that is to say, rotations like the rotation of the Earth around its polar axis. Once the primordial condensations had formed into a compact state, their rotation period must have fallen below seven hours, and as Lyttleton showed in 1938, a solid planet with such a rotation period must break into two pieces, one considerably bigger than the other. The great planets must be the main chunks arising from these processes of break-up. Now in the break-up it is also to be expected that a number of comparatively small blobs become detached from the main bodies as they separate from each other. For the most part, these blobs remained circling around the great planets—these are the satellites. But a few of the larger blobs seem to have escaped, and these are the five small planets—Venus, Mercury, Mars, Pluto, and the Earth. Very probably the Moon was an adjacent blob that became detached along with the Earth. So, to sum up, there were a number of big primordial planets that broke up about 2,500,000,000 years ago, and one of the bits of the debris was our Earth and another the Moon.

This picture of the way the Earth came into being is I think very important to our studies of the interior of the Earth. It affects our views on the probable temperature of the deep interior, suggesting that it may be much less than was formerly believed. It provides interesting possibilities regarding the origin of the Earth's magnetism. It leads to a plausible explanation of the origin of the surface rocks. For the Earth must have originally moved in a highly elliptic orbit that took it into the inner parts of the gaseous disk. Here the material had not been entirely swept up by the primordial condensations, which were formed much farther out from the Sun. So the Earth moved through a medium consisting partly of gas and partly of comparatively small solid bodies. This had two effects; one was to round up the Earth's motion into a nearly circular orbit that lies well inside the orbits of the great planets, and the other was that it modified the surface feature of the Earth through the acquisition of various gases and solid bodies. The solid layers known as the Earth's crust may well have originated in this way. In particular, it is possible that the Earth acquired its radioactive materials during this final stage. Among the gases acquired were probably nitrogen, water, oxygen, and carbon-dioxide. The histories of Venus, Mercury, and Mars must have been somewhat different because their orbits took them through different parts of the disk. In particular, Venus seems to have obtained little or no water but very large quantities of carbon-dioxide and also possibly nitrogen. Mars, on the other hand, obtained carbon-dioxide and water but not so much water as the Earth. The fate of Pluto we do not know.

PRODUCTS OF WATER MOLECULES

There is a further general point that we must not overlook. As the Sun and the disk moved together around the Galaxy, interstellar hydrogen must have impinged on the disk and become added to the material. Since the whole condensation process took as long as 1,000,000,000 years, the amount of hydrogen that was picked up in this way must have made quite an appreciable addition to the gas in the disk. Much of the hydrogen would probably combine with oxygen and so water would be formed. In fact, water would eventually become a fairly common molecule. It is likely that the shells of ice that we believe to be a feature of the great planets were formed out of these molecules. Other important products of the water molecules appear to be the rings of Saturn, which seem to be made of ice crystals, the five inner satellites of Saturn which seem to be large snowballs, and the oceans of the Earth.

By now we have considered the origin of both the Sun and the planets. To complete our cosmological picture I should also say a little about the probable final fate of the solar system. I will do this on the supposition that the Sun is not going to sweep up large quantities of interstellar gas by the tunnelling process I discussed last week. Then the amount of material in the Sun will remain pretty much as it is at present. On this basis the future history of the Sun during the next 50,000,000,000 years or so will follow the lines I have already described in these lectures, when I said that the Sun will grow steadily more luminous as its hydrogen supply is converted into helium, and this will go on until the oceans boil on the Earth. And I then went on to say that as the Sun grills the Earth it will swell, at first slowly and then with increasing rapidity, until it swallows the inner planets one by one: first Mercury, then Venus, and then the Earth. Mars is likely to be the last planet to suffer this fate, but it is possible that an even further extension as far as Jupiter may occur.

All this refers to a stage just before the Sun's hydrogen becomes exhausted. Once the internal hydrogen is used up energy generation through the building of helium will cease, and the Sun will then begin to collapse. Its swollen size will disappear. As it shrinks the surface will change from the dull red colour that must occur in the distended state I have just described. First the surface will warm up to a bright red, then to a white heat, and then to a fierce electric blue. Will the Sun become an exploding star? The answer to this question is, no.

When the Sun has shrunk to about the size of the Earth a new form of pressure begins to develop inside. This new pressure is important because it operates without a high temperature being necessary. When it comes into action it will allow the Sun to cool off without any further collapse being necessary. I cannot explain here exactly why this should be possible, but it is what will happen for a star like the Sun. In such a star as the one-time parent of our planets, cooling off could not occur, because as Stoner of Leeds and Chandrasekhar of Chicago have shown, the new form of pressure I have just mentioned is not powerful enough to prevent the collapse of really massive stars.

To continue with our story of the eventual fate of the solar system. Once the Sun starts cooling off, the escape of radiation from its surface into surrounding space will reduce the temperature in the interior. After about 500,000,000 years the steely blue colour of the surface will change to white. The Sun will then be what is called a white dwarf. With the further passage of aeons greater than the present ages of the stars, the surface will cool to a dull red, and then after the elapse of a still greater span of time the light will go out altogether and the Sun will be a black dwarf that moves through space accompanied by its retinue of unlit planets—that is to say, accompanied by those planets that it had not consumed at an earlier stage. As I have referred to the white dwarfs, this is the proper time to mention the fate of the stellar remnant that was left over from the Sun's one-time companion. This remnant of the parent of our planets is now almost certainly a faint white dwarf lying somewhere in the Galaxy. Once it had rid itself of most of the material in the supernova explosion, this remnant must have been able to cool off, and it is now going through its final evolution to a black dwarf.

LIFE ON OTHER PLANETS

By now we have covered enough ground for us to refer back to the end of my first talk when I said that there are about 1,000,000 planetary systems in the

Milky Way in which life may exist. I should like now to tell you how I made this estimate. No supernova outburst is visible in the Milky Way at the present time. But the gases hurled into space by the supernova observed by the Chinese in A.D. 1054 actually can be seen. It was these gases that furnished Baade and Minkowski with the information I mentioned above. Since A.D. 1054 two other supernovae have also blazed out in the Milky Way, one in 1572 and the other in 1604.

This suggests that on the average one supernova occurs every two or three hundred years. At this rate there must have been more than 10,000,000 supernova explosions since the oldest stars were born—this was about 4,000,000,000 years ago, as I explained last week. Now something like a half of all these supernovae must have been components in binary systems, and must accordingly have given birth to planets in exactly the way we have discussed here. So in the past, nearly 10,000,000 planetary systems, each one being similar to the solar system in the essential features of its construction, must have been formed in the Milky Way.

Next we ask what proportion of these systems would contain a planet on which the physical conditions were suitable for the support of life. I estimate for this about one planetary system in ten, which gives me a final total of 1,000,000 possible abodes of life within the Milky Way. I will admit that the last bit of this calculation is approximate. But even when full allowance is made for all the uncertainties I do not think that the final total could be less than 100,000. Our next question is: Will living creatures arise on every planet where favourable physical conditions occur?

No certain answer can be given to this, but those best qualified to judge the matter, the biologists, seem to think that life would in fact arise wherever conditions were able to support it. Accepting this, we can proceed with greater assurance. The extremely powerful process of natural selection would come into operation and would shape the evolution of life on each of these distant planets. Would creatures arise having some sort of similarity with those on the Earth? In a broadcast last summer the distinguished biologist, C. D. Darlington, showed that this is by no means as unlikely as it seems at first sight. After remarking on the inevitability of the way life has evolved on the Earth, Darlington pointed out that the human body is not a rickety accidental structure, but an extremely fine product of superb construction. And, as you will realize, if a creature like man is a sound proposition on the Earth, then similar creatures, built on a similar plan, would also be well fitted to an existence on other planets. Darlington concludes: 'There are such very great advantages in walking on two legs, in carrying one's brain in one's head, in having two eyes on the same eminence at a height of five or six feet, that we might as well take quite seriously the possibility of a pseudo man and a pseudo woman with some physical resemblance to ourselves. . . .'

Let us end by putting all this in another way. I have often seen it stated that our situation on the Earth is providential. The argument goes like this. It is providential that the Earth is of the right size and is at the right distance from the Sun. It is providential that the Sun radiates the right kind of light and heat. It is providential that the right chemical substances occur on the Earth. A long list of this sort of statement could be compiled, and to some people it looks as if there is indeed something very strange and odd about our particular home in the Universe. But I think that this outlook arises from a misunderstanding of the situation. Because if everything was not just right we should not be here. We should be somewhere else.

GALAXY IN ANDROMEDA

The bright dots covering the photograph are foreground stars belonging to our own Galaxy.

(*Mount Wilson and Palomar Observatories*)

MAN'S PLACE IN THE EXPANDING UNIVERSE

AT the risk of seeming a little repetitive I should like to begin with a few remarks about my previous talks. One of the things I have been trying to do is to break up our survey of the Universe into distinct parts. We started with the Sun and our system of planets. To get an idea of the size of this system we took a model of our solar system with the Sun represented by a ball about six inches in diameter. In spite of this enormous reduction of scale our model would still cover the area of a small town. On the same scale the Earth has to be represented by a speck of dust, and the nearest stars are 2,000 miles away. So it is quite unwieldy to use this model to describe the positions of even the closest stars.

Some other means had to be found to get to grips with the distances of the stars in the Milky Way. Choosing light as our measure of distance, we saw that light takes several years to travel to us from nearby stars, and that many of the stars in the Milky Way are as much as 1,000 light years away. But the Milky Way is only a small bit of a great disk-shaped system of gas and stars that is turning in space like a great wheel. The diameter of the disk is about 60,000 light years. This distance is so colossal that there has only been time for the disk to turn round about twenty times since the oldest stars were born—about 4,000,000,000 years ago. And this is in spite of the tremendous speed of nearly 1,000,000 miles an hour at which the outer parts of the disk are moving. We also saw that the Sun and our planets lie together near the edge of our Galaxy, as this huge disk is called.

Vast Numbers of Other Galaxies

To-night we shall go out into the depths of space far beyond the confines of our own Galaxy. Look out at the heavens on a clear night; if you want a really impressive sight do so from a steep mountain-side or from a ship at sea. As I have said before, by looking at any part of the sky that is distant from the Milky Way you can see right out of the disk that forms our Galaxy. What lies out there? Not just scattered stars by themselves, but in every direction space is strewn with whole galaxies, each one like our own. Most of these other galaxies—or extra-galactic nebulae as astronomers often call them—are too faint to be seen with the naked eye, but vast numbers of them can be observed with a powerful telescope. When I say that these other galaxies are similar to our Galaxy I do not mean that they are exactly alike. Some are much smaller than ours, others are not disk-shaped but nearly spherical in form. The basic similarity is that they are all enormous clouds of gas and stars, each one with anything from 100,000,000 to 10,000,000,000 or so members.

Although most of the other galaxies are different from ours, it is important to realize that some of them are indeed very like our Galaxy even so far as details are concerned. By good fortune one of the nearest of them, only about 700,000 light years away, seems to be practically a twin of our Galaxy. You can see it for yourself by looking in the constellation of Andromeda. With the naked eye it appears as a vague blur, but with a powerful telescope it shows up as one of the most impressive of all astronomical objects. On a good photograph of it you can easily pick out places where there are great clouds of dust. These clouds are just the sort of thing that in our own Galaxy produces the troublesome fog I mentioned in earlier talks. It is this fog that stops us seeing more than a small bit of our Galaxy. If you want to get an idea of what our Galaxy would look like if it were seen from outside, the best way is to study this other one in Andromeda. If the truth be known I expect that there are plenty of places there where living creatures are looking out across space at our Galaxy, and who are seeing much the same spectacle as we see when we look at their galaxy.

If there were time I could say a great deal about the properties of all these other galaxies: how they are spinning round like our own; how their brightest stars are supergiants, just like those of our Galaxy; and how in those where supergiants are common, wonderful spiral patterns are found. A question that interests me very much is whether these spiral patterns are connected with the tunnelling process I discussed a fortnight ago. Then there is the way in which Mayall and Wyse of the Lick Observatory inferred that the galaxy in Andromeda contains huge tracts of gas, like the interstellar gas in our own Galaxy.

We can find also exploding stars, the ones I described

last week, in these other galaxies. In particular, supernovae are so brilliant that they show up even when
they are very far off. Now the existence of supernovae
in other galaxies has implications for our cosmology.
You will remember that last week I described the way
in which planetary systems like our own solar system
come into being, and that the basic requirement of the
process was the supernova explosion. So we can conclude, since supernovae occur in the other galaxies,
planetary systems must exist there just as in our own.
Moreover, by observing the other galaxies we get a far
better idea of the rate at which supernovae occur than
we could ever get from our Galaxy alone. A general
survey by the American observers Baade and Zwicky
has shown that on the average there is a supernova
explosion every four or five hundred years in each
galaxy. So, remembering the evidence I presented last
week, you will see that on the average each galaxy
must contain more than 1,000,000 planetary systems.

How many of these gigantic galaxies are there? Well,
they are strewn through space as far as we can see with
the most powerful telescopes. Spaced apart at an
average distance of rather more than 1,000,000 light
years, they certainly continue out to the fantastic
distance of 1,000,000,000 light years. Our telescopes
fail to penetrate further than that, so we cannot be
certain that the galaxies extend still deeper into space,
but we feel pretty sure that they do. One of the
questions we shall have to consider later is what lies
beyond the range of our most powerful instruments.
But even within the range of observation there are
about 100,000,000 galaxies. With upwards of 1,000,000
planetary systems per galaxy the combined total for

the parts of the Universe that we can see comes out at more than a hundred million million. I find myself wondering whether somewhere among them there is a cricket team that could beat the Australians.

ORIGIN OF THE GALAXIES

We now come to the important question of where this great swarm of galaxies has come from. Perhaps I should first remind you of what we said when we were discussing the origin of the stars. We saw that in the space between the stars of our Galaxy there is a tenuous gas, the interstellar gas. At one time our Galaxy was a whirling disk of gas with no stars in it. Out of the gas, clouds condensed, and then in each cloud further condensations were formed. This goes on until finally stars are born. Stars are formed in the other galaxies in exactly the same way. Not only this, but we can go further and extend the condensation idea to include the origin of the galaxies themselves. Just as the basic step in explaining the origin of the stars is the recognition that a tenuous gas pervades the space within a galaxy, so the basic step in explaining the origin of the galaxies is the recognition that a still more tenuous gas fills the *whole of space*. It is out of this general background material, as I shall call it, that the galaxies have condensed.

Here now is a question that is important for our cosmology. What is the present density of the background material? The average density is so low that a match-box would contain only about one atom. But small as this is, the total amount of the background material exceeds about a thousandfold the combined quantity of material in all the galaxies put together. This may seem surprising but it is a consequence of the fact that the galaxies occupy only a very small fraction of the whole of space. You see here the characteristic signature of the New Cosmology. We have seen that inside our Galaxy the interstellar gas outweighs the material in all the stars put together. Now we see that the background material outweighs by a large margin all the galaxies put together. And just as it is the interstellar gas that controls the situation inside our Galaxy, so it is the background material that controls the Universe as a whole. This will become increasingly clear as we go.

The degree to which the background material has to be compressed to form a galaxy is not at all comparable with the tremendous compression necessary to produce a star. This you can see by thinking of a model in which our Galaxy is represented by a penny. Then the blob of background material out of which our Galaxy condensed would be only about a foot in diameter. This incidentally is the right way to think about the Universe as a whole. If in your mind's eye you take the average galaxy to be about the size of a bee, our Galaxy, which is a good deal larger than the average, would be roughly represented in shape and size by a penny, and the average spacing of the galaxies would be about three yards, and the range of telescopic vision about a mile. So sit back and imagine a swarm of bees spaced about three yards apart and stretching away from you in all directions for a distance of about a mile. Now for each bee substitute the vast bulk of a galaxy and you have an idea of the Universe that has been revealed by the large American telescopes.

THE EXPANSION OF SPACE

Next I must introduce the idea that this colossal swarm is not static: it is expanding. There are some people who seem to think that it would be a good idea if it was static. I disagree with this idea, if only because a static universe would be very dull. To show you what I mean by this I should like to point out that the Universe is wound up in two ways—that is to say, energy can be got out of the background material in two ways. Whenever a new galaxy is formed, gravitation supplies energy. For instance, gravitation supplies the energy of the rotation that develops when a galaxy condenses out of the background material. And gravitation again supplies energy during every subsequent condensation of the interstellar gas inside a galaxy. It is because of this energy that a star becomes hot when it is born. The second source of energy lies in the atomic nature of the background material. It seems likely that this was originally pure hydrogen. This does not mean that the background material is now entirely pure hydrogen, because, as I said last week, it gets slightly adulterated by some of the material expelled by the exploding supernovae. As a source of energy hydrogen does not come into operation until high temperatures develop—and this only arises when stars condense. It is this second source of energy that is more familiar and important to us on the Earth.

Why would a Universe that was static on a large scale, that was not expanding in fact, be uninteresting? Because of the following sequence of events. Even if the Universe were static on a large scale it would not be locally static: that is to say, the background material would condense into galaxies, and after a few thousand million years this process would be completed—no background would be left. Furthermore, the gas out of which the galaxies were initially composed would condense into stars. When this stage was reached hydrogen would be steadily converted into helium. After several hundreds of thousands of millions of years this process would be everywhere completed and all the stars would evolve towards the black dwarfs I mentioned last week. So finally the whole Universe would become entirely dead. This would be the running down of the Universe that was described so graphically by Jeans.

My main purpose to-night is to explain why we get a different answer to this when we take account of the dynamic nature of the Universe. You might like to hear something about the observational evidence that the Universe is indeed in a dynamic state of expansion. It comes from spectroscopy—the study of the sort of light emitted by particular atoms and molecules. The light emitted, for example, from atoms of iron, has a characteristic pattern. Now we can recognize the pattern of iron in the light we receive from many distant galaxies. But the pattern is not quite the same as the light emitted by iron atoms in a laboratory on the Earth. There is the curious difference that the pattern from a distant galaxy is always systematically lower in pitch, or as we usually say reddened, as compared with a similar source of light on Earth. At first sight this may seem rather mysterious, but we know from common experience that this reddening is just what happens whenever we receive light from a body that is moving away from us. In fact, by measuring the degree of the reddening, we can deduce the speed with which a body is receding.

When this is applied to the galaxies in general it leads to the startling conclusion that the galaxies are moving away from us. And what is more, the greater the distance of a galaxy the faster it is receding. Every time you double the distance you double the speed of recession. The speeds come out as vast beyond all precedent. Nearby galaxies are moving outwards at several million miles an hour, whereas the most distant ones that can be seen with our biggest telescopes are receding at over 200,000,000 miles an hour. This leads us to the obvious question: If we could see galaxies lying at even greater distances, would their speeds be still vaster? Nobody seriously doubts that this would be so, which gives rise to a very curious situation that I will now describe.

Galaxies lying at only about twice the distance of the furthest ones that actually can be observed with the new telescope at Mount Palomar would be moving away from us at a speed that equalled light itself. Those at still greater distances would have speeds of recession exceeding that of light. Many people find this extremely puzzling because they have learnt from Einstein's special theory of relativity that no material body can have a speed greater than light. This is true enough in the special theory of relativity which refers to a particularly simple system of space and time. But it is not true in Einstein's general theory of relativity. The point is rather difficult, but I can do something towards making it a little clearer.

The trouble really arises from too naïve an idea of

what is meant by the speed of recession of a galaxy. Imagine a number of dots marked on a flat sheet of rubber. Now choose any dot you like as a standard of reference. Then the sheet of rubber can be stretched out in such a way that, *quite regardless of which dot you have chosen*, all the other dots move away from it. This gives you a rough idea of the way in which general relativity can be used to describe the movement of the galaxies away from each other. But there is one detail you must get right if the analogy is to be useful. You must imagine the rubber to be stretched out in such a way that the dots themselves remain of practically the same size—that is to say, as the stretching proceeds, the dots stay of about the same size whereas the distances between them increase. What it amounts to is this. The distances between the galaxies increase not so much because they move in the sense we are used to in everyday life, but because the space between them gets stretched. From this point of view we can regard the reddening of light as arising in the following way. During the time taken by light to reach us from a distant galaxy the distance across which it has to travel increases owing to the stretching effect, and it is this that produces the reddening.

The Observable Universe

Now we reach a crucial point that we must grasp if we are to understand the nature of the Universe. The further a galaxy is away from us the more its distance will increase during the time required by its light to reach us. Indeed, if it is far enough away the light never reaches us at all because its path stretches faster than the light can make progress. This is what is meant by saying that the speed of recession exceeds the velocity of light. Events occurring in a galaxy at such a distance can never be observed at all by anyone inside our Galaxy, no matter how patient the observer and no matter how powerful his telescope. All the galaxies that we actually see are ones that lie close enough for their light to reach us in spite of the stretching that is going on. But the struggle of the light against the expansion of space does show itself, as I said before, in the reddening of the light. As you will easily guess, there must be a critical intermediate case where a galaxy is at such a distance that, so to speak, the light neither gains ground nor loses it. In this case the path between us and the galaxy stretches at just such a rate as exactly compensates for the velocity of the light. It is a case, as the Red Queen remarked to Alice, of 'taking all the running you can do to keep in the same place.'

We know fairly accurately how far away a galaxy has to be for this critical case to occur. The answer is about 2,000,000,000 light years, which is only about twice as far as the distances that we expect the giant telescope at Mount Palomar to penetrate. This means that we are already observing about half as far into space as we can ever hope to do. If we built a telescope a million times as big as the one at Mount Palomar we could not even double our present range of vision. So what it amounts to is that owing to the expansion of the Universe we can never observe events that happen outside a certain quite definite finite region of space. We shall refer to this finite region as the observable universe. The word 'observable' here does not mean what we actually observe, but what we could observe if we were equipped with perfect telescopes.

THEORIES OF THE EXPANDING UNIVERSE

We must move on to consider the explanations that have been offered for this expansion of the Universe. First I will consider the older ideas—that is to say, the ideas of the nineteen-twenties and the nineteen-thirties— and then I will go on to offer my own opinion. Broadly speaking, the older ideas fall into two groups. One was that the Universe started its life a finite time ago in a single huge explosion, and that the present expansion is a relic of the violence of this explosion. This big bang idea seemed to me to be unsatisfactory even before detailed examination showed that it leads to serious difficulties. For when we look at our own Galaxy there is not the smallest sign that such an explosion ever occurred. This might not be such a cogent argument if on such theories our Galaxy were much younger than the whole Universe. But this is not so. In fact, in some of these theories there is the obvious contradiction that the Universe comes out to be younger than our astrophysical estimates of the age of our Galaxy. But the really serious difficulty arises when we try to reconcile the idea of an explosion with the requirement that the galaxies have condensed out of diffuse background material. The two concepts of explosion and condensation are obviously contradictory, and it is easy to show, if you postulate an explosion of sufficient violence to explain the expansion of the Universe, that condensations looking at all like the galaxies could never have been formed.

We come now to the second group of theories. These attempt to explain the expansion of the Universe by monkeying with the law of gravitation. The ordinary idea that two particles attract each other is only accepted if their distance apart is not too great. At really large distances, so the argument goes, the two particles repel each other instead. On this basis it can be shown that if the density of the background material is sufficiently small, expansion must occur. But once again there is a difficulty in reconciling this with the requirement that the background material must condense to form the galaxies. For once the law of gravitation has been modified in this way the tendency is for the background material to be torn apart rather than for it to condense into galaxies. Actually there is just one way in which a theory along these lines can be built so as to get round this difficulty. This is a theory due to Lemaître which was often discussed by Eddington in his popular books. But we now know that on this theory the galaxies would have to be vastly older than our astrophysical studies show them to be. So even this has to be rejected.

I should like now to approach more recent ideas by describing what would be the fate of our observable universe if any of these older theories had turned out to be correct. Every receding galaxy will eventually increase its distance from us until it passes beyond the limit of the observable universe—that is to say, they will move to a distance beyond the critical limit of about 2,000,000,000 light years that I have already mentioned. When this happens, nothing that occurs within them can ever be observed from our Galaxy. So if any of the older theories were right we should end in a seemingly empty universe, or at any rate in a universe that was empty apart perhaps from one or two very close galaxies that became attached to our Galaxy as satellites. Nor would this situation take very long to develop. Only about 10,000,000,000 years—that is to say, about a fifth of the lifetime of the Sun—would be needed to empty the sky of the 100,000,000 or so galaxies that we can now observe.

Although I think there is no doubt that every galaxy we now observe to be receding from us will in about 10,000,000,000 years have passed entirely beyond the limit of vision of an observer in our Galaxy, yet I think that such an observer would still be able to see about the same number of galaxies as we do now. By this I mean that new galaxies will have condensed out of the background material at just about the rate necessary to compensate for those that are being lost as a consequence of their passing beyond our observable universe. At first sight it might be thought that this could not go on indefinitely because the material forming the background would ultimately become exhausted. But again I do not believe that this is so, for it seems likely that new material is constantly being created so as to maintain a constant density in the background material. So we have a situation in which the loss of galaxies, through the expansion of the Universe, is compensated by the condensation of new galaxies, and this can continue indefinitely.

The idea that matter is created continuously represents our ultimate goal in this series of lectures. The idea in itself is not new. I know of references to the continuous creation of matter that go back more than twenty years, and I have no doubt that a close inquiry would show that the idea, in its vaguest form, goes back very much further than that. What is new is it has now been found possible to put a hitherto vague idea in a precise mathematical form. It is only when this has been done that the consequences of any physical idea can be worked out and its scientific value assessed. I should perhaps explain that besides my personal views which I shall now be putting forward there are two other lines of thought on this matter. One is due to the German scientist P. Jordan, whose views differ from my own by so wide a gulf that it would be too wide a digression to discuss them. The other line of attack has been due to the Cambridge scientists H. Bondi and T. Gold who, although using quite a different form of argument to the one I adopted, have reached conclusions almost identical with those I am now going to discuss.

From time to time people ask where the created material comes from. Well, it does not come from anywhere. Material simply appears—it is created. At one time the various atoms composing the material do not exist and at a later time they do. This may seem a very strange idea and I agree that it is, but in science it does not matter how strange an idea may seem so long as it works—that is to say, so long as the idea can be expressed in a precise form and so long as its consequences are found to be in agreement with observation. In any case, the whole idea of creation is queer. In the older theories all the material in the Universe is supposed to have appeared at one instant of time, the whole creation process taking the form of one big bang. For myself I find this idea very much queerer than continuous creation.

Perhaps you may think that the whole question of the creation of the Universe could be avoided in some way. But this is not so. To avoid the issue of creation it would be necessary for all the material of the Universe to be infinitely old, and this it cannot be. For if this were so, there could be no hydrogen left in the Universe. Hydrogen is being steadily converted into helium and the other elements throughout the Universe and this conversion is a one-way process—that is to say, hydrogen cannot be produced in any appreciable quantity through the breakdown of the other elements. How comes it then that the Universe consists almost entirely of hydrogen? If matter were infinitely old this would be quite impossible. So we see that the Universe being what it is, the creation issue simply cannot be dodged. And I think that of all the various possibilities that have been suggested, continuous creation is easily the most satisfactory.

Now what are the consequences of continuous creation? Perhaps the most surprising result of the mathematical theory is that the average density of the background material must stay constant. To achieve this only a very slow creation rate is necessary. The new material does not appear in a concentrated form in small localized regions but is spread throughout the whole of space. The average rate of appearance amounts to no more than the creation of one atom in the course of about a year in a volume equal to St. Paul's Cathedral. As you will realize, it would be quite impossible to detect such a rate of creation by direct experiment. But although this seems such a slow rate when judged by ordinary ideas, it is not small when you consider that it is happening everywhere in space. The total rate for the observable universe alone is about a hundred million, million, million, million, million tons per second. Do not let this surprise you because, as I have said, the volume of the observable universe is very large. It is this creation that drives the Universe. The new material produces an outward pressure that leads to the steady expansion. But it does much more than that. With continuous creation the apparent contradiction between the expansion of the Universe and the requirement that the background material shall be able to condense into galaxies is completely overcome. For it can be shown that once an irregularity occurs in the background material a galaxy must eventually be formed. Such irregularities are constantly being produced through the gravitational action of the galaxies themselves. So the background material must give a steady supply of new galaxies. Moreover, the created material also supplies unending quantities of atomic energy, for by arranging that newly created material is composed of hydrogen we explain why in spite of the fact that hydrogen is being consumed in huge quantities in the stars the Universe is nevertheless observed to be overwhelmingly composed of it.

WHAT LIES BEYOND?

It is now time for us to consider the question as to what lies beyond the observable part of the Universe. In the first place, does this question have any meaning? According to the theory it does. Theory requires the galaxies to go on for ever, even though we cannot see them. That is to say, the galaxies are expanding out into an infinite space. There is no end to it all. And what is more, apart from the possibility of there being a few freak galaxies, one bit of this infinite space will behave in the same way as any other bit.

H

The same thing applies to time. You will have noticed that in these lectures I have used the concepts of space and time as if they could be treated separately. According to the relativity theory this is a dangerous thing to do. But it so happens that it can be done with impunity in our Universe, although it is easy to imagine other universes where it could not be done. What I mean by this is that a division between space and time can be made and this division can be used throughout the whole of our Universe. This is a very important and special property of our Universe, which I think it is important to take into account in forming the equations that decide the way in which matter is created.

Let us suppose that a film is made from any space position in the Universe. To make the film let a still picture be taken at each instant of time. This, by the way, is what we are doing in our astronomical observations. We are actually taking the picture of the Universe at one instant of time—the present. Next, let all the stills be run together so as to form a continuous film. What would the film look like? Well, galaxies would be observed to be continually condensing out of the background material. The general expansion of the whole system would be clear, but though the galaxies seemed to be moving away from us there would be a curious sameness about the film. It would be only in the details of each galaxy that changes would be seen. The overall picture would stay the same because of the compensation whereby the galaxies that were constantly disappearing through the expansion of the Universe were replaced by newly forming galaxies. A casual observer who went to sleep during the showing of the film would find it difficult to see much change when he awoke. How long would our film show go on? It would go on for ever.

There is a complement to this result that we can see by running our film backwards. Then new galaxies would appear at the outer fringes of our picture as faint objects that come gradually closer to us. For if the film were run backwards the Universe would appear to contract. The galaxies would come closer and closer to us until they evaporated before our eyes. First the stars of a galaxy would evaporate back into the gas from which they were formed. Then the gas in the galaxy would evaporate back into the general background from which it had condensed. The background material itself would stay of constant density, not through matter being created, but through matter disappearing. How far could we run our hypothetical film back into the past? Again according to the theory, for ever. After we had run backwards for about 5,000,000,000 years our Galaxy itself would disappear before our eyes. But although important details like this would no doubt be of great interest to us there would again be a general sameness about the whole proceeding. Whether we run the film backwards or forwards the large-scale features of the Universe remain unchanged.

So we see that no large-scale changes in the Universe can be expected to take place in the future. But individual galaxies will change. What is the history of our own Galaxy going to be? This issue cannot be decided by observation because none of the galaxies that we observe can be more than about 10,000,000,000 years old. The reason for this is that a new galaxy condensing close by our Galaxy moves away from us and will pass out of the observable region of space in only about 10,000,000,000 years. So we have to decide the ultimate fate of our Galaxy again from theory. It will become steadily more massive as more and more of the background material gets pulled into it. After about 10,000,000,000 years it is likely that our Galaxy will have succeeded in gathering quite a cloud of gas and satellite bodies. Where this will ultimately lead is difficult to say with any precision. The distant future of the Galaxy is to some extent bound up with an investigation made about thirty years ago by Schwarzschild, who found that very strange things happen when a body grows particularly massive. It becomes difficult, for instance, for light emitted by the body ever to get out into surrounding space. When this stage is reached further growth is likely to be strongly inhibited. Just what it would then be like to live in our Galaxy I should very much like to know.

To conclude the more astronomical part of this lecture, I should like to stress that so far as the Universe as a whole is concerned the essential difference made by the idea of continuous creation of matter is this. Without continuous creation the Universe must evolve towards a dead state in which all the matter is condensed into a vast number of dead stars. The details of the way this happens are different in the different theories that have been put forward, but the outcome is always the same. With continuous creation, on the other hand, the Universe has an infinite future in which all its present very large-scale features will be preserved.

Looking to the Future

I come now to an entirely different class of question, which, to be quite frank, I should really prefer to avoid if I could. But I am sure you would hardly wish me to end without saying something about how the New Cosmology affects me personally. So with the clear understanding that what I am now going to say has no agreed basis among scientists, but represents my own personal views, I shall try to sum up the general philosophic issues that seem to me to come out of our survey of the Universe.

It is my view that man's unguided imagination could never have chanced on such a structure as I have put before you in these talks. No literary genius could have invented a story one-hundredth part as fantastic as the sober facts that have been unearthed by astronomical science. You need only compare our inquiry into the nature of the Universe with the tales of such acknowledged masters as Jules Verne and H. G. Wells to see that fact outweighs fiction by an enormous margin. One is naturally led to wonder what the impact of the New Cosmology would have been on a man like Newton who would have been able to take it in, details and all, in one clean sweep. I think that Newton would have been quite unprepared for any such revelation, and that it would have had a shattering effect on him.

Is it likely that any astonishing new developments are lying in wait for us? Is it possible that the cosmology of 500 years hence will extend as far beyond our present beliefs as our cosmology goes beyond that of Newton? It may surprise you to hear that I doubt whether this will be so. I am prepared to believe that there will be many advances in the detailed understanding of matters that still baffle us. Of the larger issues I expect a considerable improvement in the theory of the expanding universe. Already it is fairly clear that the theory of relativity is not an ideal tool for dealing with this problem. Continuous creation I expect to play an important role in the theories of the future. Indeed, I expect that much will be learned about continuous creation, especially in its connection with atomic physics. But by and large I think that our present picture will turn out to bear an appreciable resemblance to the cosmologies of the future. If this should appear presumptuous to you, I think you should consider what I said earlier about the observable region of the Universe. As you will remember, even with a perfect telescope, we could penetrate only about twice as far into space as the new telescope at Mount Palomar. This means that there are no new fields to be opened up by the telescopes of the future, and this is a point of no small importance in our cosmology.

In all this I have assumed that progress will be made in the future. It is quite on the cards that astronomy may go backwards, as, for instance, Greek astronomy went backwards after the time of Hipparchus. And in saying this I am not thinking about an atomic war destroying civilization, but about the increasing tendency to rivet scientific inquiry in fetters. Secrecy, nationalism, the Marxist ideology, these are some of the things that are threatening to choke the life out of science. You may possibly think that this might be a good thing, as we have obviously had quite enough of atom bombs, disease-spreading bacteria, and radioactive poisons to last us for a long time. But this is not the way in which it works. What will happen if science declines is that there will be more work, not less, on the comparatively easy problems of destruction. It will be the real science, where the adversary is not man but the Universe itself, that will suffer. I should like to think that, in saying all this, I was being an alarmist, but unfortunately it seems that almost every development during the last fifteen years has taken the world along the wrong road. I think that it may have been the recognition of this that recently led the biologists to protest so strongly over the Lysenko scandal.

A PERSONAL VIEW

Next I come to the question that has been lying in ambush ever since I started these talks. What is man's place in the Universe? I should like to make a start on this momentous issue by considering the view of the out-and-out materialists. The appeal of their argument is based on simplicity. The Universe is here, they say, so let us take it for granted. Then the Earth and other planets must arise in the way I discussed last week. On a suitably favoured planet like the Earth, life would be very likely to arise, and once it had started, so the argument goes on, only the biological processes of mutation and natural selection are needed to produce living creatures as we know them. Such creatures are no more than ingenious machines that have evolved as strange by-products in an odd corner of the Universe. No important connection exists, so the argument concludes, between these machines and the Universe as a whole, and this explains why all attempts by the machines themselves to find such a connection have failed.

Most people object to this argument for the not very good reason that they do not like to think of themselves as machines. But taking the argument at its face value, I see no point that can actually be disproved, except the claim of simplicity. The outlook of the materialists is not simple, it is really very complicated. For instance, it is definitely up to the materialists to explain how consciousness has evolved in the human machine, exactly how your consciousness and mine can be squared with the machine idea. I can see that a sort of robot machine might be produced by normal biological processes, but exactly how is a machine produced that can think about itself and the Universe as a whole? At just what stage in the evolution of living creatures did individual consciousness arise? I do not say that questions such as these are unanswerable, but I do say that it will not be simple to answer them.

But all this is a minor issue compared with what seems to me to be the real objection to the outlook of the materialists. The apparent simplicity, such as it is, of their case is only achieved by taking the existence of the Universe for granted. For myself there is a great deal more about the Universe that I should like to know. Why is the Universe as it is and not something else? Why is the Universe here at all? It is true that at present we have no clue to the answers to questions such as these, and it may be that the materialists are right in saying that no meaning can be attached to them. But throughout the history of science people have been asserting that such and such an issue is inherently beyond the scope of reasoned inquiry, and time after time they have been proved wrong. Two thousand years ago it would have been thought quite impossible to investigate the nature of the Universe to the extent I have been describing it to you in these lectures. And I dare say that you yourself would have said, not so very long ago, that it was impossible to learn anything about the way the Universe is created. All experience teaches us that no one has yet asked too much. How then can we accept the argument of the materialists, when the essence of their game lies in throwing up the sponge?

And now I should like to give some consideration to contemporary religious beliefs. There is a good deal of cosmology in the Bible. My impression of it is that it is a remarkable conception, considering the time when it was written. But I think it can hardly be denied that the cosmology of the ancient Hebrews is only the merest daub compared with the sweeping grandeur of the picture revealed by modern science. This leads me to ask the question: Is it in any way reasonable to suppose that it was given to the Hebrews to understand mysteries far deeper than anything I can comprehend, when it is quite clear that they were completely ignorant of many matters that seem commonplace to me? No, it seems to me that religion is but a blind attempt to find an escape from the truly dreadful situation in which we find ourselves. Here we are in this wholly fantastic Universe with scarcely a clue as to whether our existence has any real significance. No wonder then that many people feel the need for some belief that gives them a sense of security, and no wonder that they become very angry with people like me who say that this security is illusory. But I do not like the situation any better than they do. The difference is that I cannot see how the smallest advantage is to be gained from deceiving myself. We are in rather the situation of a man in a desperate, difficult position on a steep mountain. A materialist is like a man who becomes crag-fast and keeps on shouting: 'I'm safe, I'm safe!' The religious person is like a man who goes to the other extreme and rushes up the first route that shows the faintest hope of escape, and who is entirely reckless of the yawning precipices that lie below him.

I will illustrate all this by saying what I think about perhaps the most inscrutable question of all: Do our minds have any continued existence after death? To make any progress with this question it is necessary to understand what our minds are. If we knew this with any precision then I have no doubt we should be well on the way to getting a satisfactory answer. But at the moment we have only got the vaguest of ideas on this. One thing, however, seems clear—that mind, if it exists in the religious sense, must have some physical connections. That is to say, if the something we call mind does survive death then this something must be capable of physical detection. Both modern physiology and religion seem to be agreed about this point. Towards the end of Handel's *Messiah* there are words that run something like this: 'Although I die, yet in my flesh shall I see God.' Notice the words 'in my flesh.' In any case, survival after death would be meaningless and unthinkable without some interaction with the physical world. Besides, if the mind were without physical connections, why is it that the mind is so intimately associated with the body?

Here then is a positive way to attack the problem. When a person dies, does a mind that is physically detectable survive? Eventually it should be possible to decide this question. Some people might say that such a survival would have been detected already if it existed. But I do not think there is anything to warrant this belief. If it were not for the fact that we happen to live near a large body—namely, the Earth—I doubt whether gravitation itself would so far have been discovered by experiments in the laboratory. It is quite on the cards that there are new and important physical relationships still to be revealed by scientific investigation.

I should like to end by discussing in a little more detail the beliefs of the Christians as I see them myself. In their anxiety to avoid the notion that death is the complete end of our existence, they suggest what is to me an equally horrible alternative. If I were given the choice of how long I should like to live with my present physical and mental equipment, I should decide on a good deal more than seventy years. But I doubt whether I should be wise to decide on more than 300 years. Already I am very much aware of my own limitations and I think that 300 years is as long as I should like to put up with them. Now what the Christians offer me is an eternity of frustration. And it is no good their trying to mitigate the situation by saying that sooner or later my limitations would be removed, because this could not be done without altering *me*. It strikes me as very curious that the Christians have so little to say about how they propose eternity should be spent.

Perhaps I had better end by saying how I should

arrange matters if it were my decision to make. It seems to me that the greatest lesson of adult life is that one's own consciousness is not enough. What one of us would not like to share the consciousness of half a dozen chosen individuals? What writer would not like to share the consciousness of Shakespeare? What musician that of Beethoven or Mozart? What mathematician that of Gauss? What I would choose would be an evolution of life whereby the essence of each of us becomes welded together into some vastly larger and more potent structure. I think such a dynamic evolution would be more in keeping with the grandeur of the physical Universe than the static picture offered by formal religion.

What is the chance of such an idea being right? Well, if there is one important result that comes out of our inquiry into the nature of the Universe it is this: when by patient inquiry we learn the answer to any problem we always find, both as a whole and in detail, that the answer thus revealed is finer in concept and design than anything we could ever have arrived at by a random guess. And this, I believe, will be the same for the deeper issues we have just been discussing. I think that all our present guesses are likely to prove but a very pale shadow of the real thing; and it is on this note that I must now finish. Perhaps the most majestic feature of our whole existence is that while our intelligences are powerful enough to penetrate deeply into the evolution of this quite incredible Universe, we still have not the smallest clue to our own fate.

NOTES

page 11

It is no argument against the bombardment theory that many missiles must have struck the lunar surface at highly oblique angles. Whether a missile strikes obliquely or not the resulting crater must be approximately circular in form. For, just as the damage done by an ordinary bomb is due to the heat produced in the explosion and not to the metal casing of the bomb, so the blast responsible for producing a crater arises from the heat released in the impact of the missile with the lunar surface. This depends on the velocity and size of the missile but not on its direction of flight.

The white streaks that are observed to radiate from a crater (for example, from the crater Tycho) are probably due to streams of dust that were shot out in the blast that formed the crater.

page 11, *lines* 28 *and* 29

'Day' and 'night' here refer, of course, to the lunar day and the lunar night.

page 12, *line* 12

'. . . the Moon would be set in rotation again . . .'—as seen by an observer on the Earth.

page 12, *line* 19

It is known that the missile responsible for the crater in Arizona struck the Earth obliquely. The crater is approximately circular.

page 18, *line* 6

Dust clouds cannot occur unless there is a gaseous atmosphere to support them. Thus the carbon dioxide atmosphere is an essential feature of this explanation. Similar dust clouds cannot occur on the Moon or on Mercury, which do not have gaseous atmospheres.

page 18, *second paragraph*

The second of these suggestions seems the more probable.

page 36, *line* 22

'. . . the combined abundance of hydrogen is only about a third of that of the other elements.' This was taken from a review given by R. Wildt in 1947. An investigation in the last few months by W. H. Ramsey gives a higher proportion of hydrogen. But even so there remains a large discrepancy between the composition of the Sun and of the planets.

page 48, *line* 27

'Tiny particles of dust condense out of the iron atoms. . . .' Iron atoms are here considered as an example. Dust also condenses from other heavy elements.

page 50, *line* 27

Strictly, the speed of 1,000 miles an hour around the polar axis of the Earth applies only to places near the Equator. In England the speed is rather more than 500 miles an hour.

page 56, *line* 29

5,000,000 miles is more than five times the diameter of the Sun.

page 58, *line* 2

A warm damp climate at the poles cannot be explained by changes in the inclination of the Earth's polar axis. No forces exist that are large enough to produce such changes. Nor do I think any plausibility can be attached to Wegener's idea of continental drift (that is, that the surface rocks move about over large distances). Once again there are no forces capable of producing such a drift.

page 71, *line* 25

Compare with note referring to page 36, line 22.

page 76, *line* 10

This calculation is for the case in which the companion star has ten times the Sun's mass. The case here considered is a representative example.

page 88, *second paragraph*

For confirmatory evidence on the frequency of supernovae see page 94.

page 96, *line* 33

These discoveries were mainly due to Hubble.

page 97, *line* 23

This adulteration of the background material when taken in conjunction with the theory of continuous creation means that the galaxies are not composed of pure hydrogen at the time they are formed. This detail has great technical importance.

page 99, *line* 8

This is the famous Hubble-Humason velocity-distance relation.

page 107, *line* 29

The difference between the infinite space of the continuous creation theory and the finite spaces of some of the older theories can be understood in the following way. A theory giving a finite space may be thought of in terms of a balloon of finite radius that is being blown up. The radius of the balloon is a measure of *time*, and its surface represents *space*. The galaxies must be thought of as confined to this surface, which must be imagined to have three dimensions instead of the usual two. This is not so difficult as it may seem. We are all familiar with pictures in perspective—pictures where artists have represented three-dimensional scenes on two-dimensional canvases. So it is not really difficult to imagine the three dimensions of space as being confined to the surface of a balloon. Increasing time has the effect of blowing up the balloon, and as its surface stretches the galaxies recede from one another. The galaxies themselves do not stretch.

An infinite space corresponds to the case where the radius of the balloon tends to infinity. In this case any finite portion of the surface can be thought of as a portion of a flat sheet, and the blowing up of the balloon has the effect of stretching the flat sheet (as is described on page 100).

page 108, *first paragraph*

See previous note.

page 109

It is a trivial consequence that the total quantity of energy observed at any one time must be equal to the quantity observed at any other time. This is the correct statement of the conservation of energy. Thus continuous creation does not lead to non-conservation of energy. The reverse is the case, for without continuous creation the total energy observed must decrease with time in an expanding universe.

page 116, *line* 25

Several authorities on religious doctrine have assured me that this statement is incorrect and that Christians do imagine an existence without physical connections. If this is so, then Christians must be endowed with a faculty not possessed by others. I would go so far as to say that it is impossible to write half a dozen meaningful sentences concerning such an existence that do not involve some reference to the physical world.